Christian Löw

Energy Landscapes of Protein Folding: From Structure to Function

Christian Löw

Energy Landscapes of Protein Folding: From Structure to Function

Structure and Function of Intermediate States of Ankyrin Repeat Proteins

Südwestdeutscher Verlag für Hochschulschriften

Impressum/Imprint (nur für Deutschland/ only for Germany)
Bibliografische Information der Deutschen Nationalbibliothek: Die Deutsche Nationalbibliothek verzeichnet diese Publikation in der Deutschen Nationalbibliografie; detaillierte bibliografische Daten sind im Internet über http://dnb.d-nb.de abrufbar.

Alle in diesem Buch genannten Marken und Produktnamen unterliegen warenzeichen-, marken- oder patentrechtlichem Schutz bzw. sind Warenzeichen oder eingetragene Warenzeichen der jeweiligen Inhaber. Die Wiedergabe von Marken, Produktnamen, Gebrauchsnamen, Handelsnamen, Warenbezeichnungen u.s.w. in diesem Werk berechtigt auch ohne besondere Kennzeichnung nicht zu der Annahme, dass solche Namen im Sinne der Warenzeichen- und Markenschutzgesetzgebung als frei zu betrachten wären und daher von jedermann benutzt werden dürften.

Verlag: Südwestdeutscher Verlag für Hochschulschriften Aktiengesellschaft & Co. KG
Dudweiler Landstr. 99, 66123 Saarbrücken, Deutschland
Telefon +49 681 37 20 271-1, Telefax +49 681 37 20 271-0
Email: info@svh-verlag.de
Zugl.: Halle(Salle), Universität Halle-Wittenberg, Diss., 2008

Herstellung in Deutschland:
Schaltungsdienst Lange o.H.G., Berlin
Books on Demand GmbH, Norderstedt
Reha GmbH, Saarbrücken
Amazon Distribution GmbH, Leipzig
ISBN: 978-3-8381-1108-7

Imprint (only for USA, GB)
Bibliographic information published by the Deutsche Nationalbibliothek: The Deutsche Nationalbibliothek lists this publication in the Deutsche Nationalbibliografie; detailed bibliographic data are available in the Internet at http://dnb.d-nb.de.

Any brand names and product names mentioned in this book are subject to trademark, brand or patent protection and are trademarks or registered trademarks of their respective holders. The use of brand names, product names, common names, trade names, product descriptions etc. even without a particular marking in this works is in no way to be construed to mean that such names may be regarded as unrestricted in respect of trademark and brand protection legislation and could thus be used by anyone.

Publisher: Südwestdeutscher Verlag für Hochschulschriften Aktiengesellschaft & Co. KG
Dudweiler Landstr. 99, 66123 Saarbrücken, Germany
Phone +49 681 37 20 271-1, Fax +49 681 37 20 271-0
Email: info@svh-verlag.de

Printed in the U.S.A.
Printed in the U.K. by (see last page)
ISBN: 978-3-8381-1108-7

Copyright © 2010 by the author and Südwestdeutscher Verlag für Hochschulschriften Aktiengesellschaft & Co. KG and licensors
All rights reserved. Saarbrücken 2010

Energy Landscapes of Protein Folding: From Structure to Function

DISSERTATION
ZUR ERLANGUNG DES AKADEMISCHEN GRADES
doctor rerum naturalium (Dr. rer. nat.)

vorgelegt der

NATURWISSENSCHAFTLICHEN FAKULTÄT II – CHEMIE UND PHYSIK
DER MARTIN-LUTHER-UNIVERSITÄT HALLE-WITTENBERG

von Diplom-Biochemiker Christian Löw
geboren am 03.06.1979 in Straubing

Halle (Saale) 2008

Diese Arbeit wurde von Januar 2004 bis Mai 2005 am Lehrstuhl für Biochemie der Universität Bayreuth und von Juni 2005 bis Februar 2008 am Lehrstuhl für Biophysik an der Universität Halle-Wittenberg unter der Leitung von Prof. Dr. Jochen Balbach angefertigt.

Promotionsgesuch eingereicht: Juni 2008

Tag des wissenschaftlichen Kolloquiums: 13. November 2008

Prüfungsausschuss:

Prof. Dr. Jochen Balbach (Erster Gutachter)
Prof. Dr. Jochen Reinstein (Zweiter Gutachter)
JProf. Dr. Kai Tittmann (Dritter Gutachter)

Mitglieder der Prüfungskomission:

Prof. Dr. Kay Saalwaechter
JProf. Dr. Jan Kantelhardt
PD Dr. Hartmut Leipner
Prof. Dr. Milton T. Stubbs
Prof. Dr. Alfred Blume

"Enthusiasm is the most beautiful word on earth"

(Christian Morgenstern)

Contents

Abstract (English)	1
Kurzfassung (German)	3
1. Introduction	7
1.1 New View on Protein Folding	7
1.2 Protein Folding in the Cell - Chaperones	11
1.3 Repeat Proteins	12
1.4 Structure and Classification of Repeat Proteins	13
1.5 Ankyrin Repeats (ARs)	15
1.6 BAR Domains and Membrane Curvature	16
1.7 Purpose of This Thesis	17
2. Summary and Discussions	19
2.1 Protein Production	19
2.2 Folding Mechanism of CDK Inhibitor $p19^{INK4d}$	21
2.3 $P19^{INK4d}$ Between Native and Partially Folded State	28
2.4 Ankyrin Repeat Proteins of Archaea	32
2.5 The Yin and Yang of Repeat Protein Folding	37
2.6 SlyD – Prolyl Isomerase and Folding Helper	39
2.7 Bringing Your Curves to the BAR	43
3. Abbreviations	49
4. References	52
5. List of Publications	60
6. Presentation of Personal Contribution	62
7. Subprojects	63
7.1 Subproject A	63
7.2 Subproject B	81
7.3 Subproject C	103
7.4 Subproject D	121
7.5 Subproject E	147
Danksagung	181

Abstract

Proteins are the vital molecules in living cells and important targets for pharmaceutical and biotechnological applications. To function they need a defined three dimensional structure. All the information necessary for a protein to achieve this conformation is encoded in its amino acid sequence. To understand the process by which a polypeptide chain folds into its correct three-dimensional structure (the so-called "protein folding reaction") is an essential element in structural biology. Modern equilibrium and kinetic spectroscopic methods provide a powerful tool to elucidate the basis of such protein folding reactions.

In this work, protein folding studies of ankyrin repeat (AR) proteins were one of the major focus. The AR is a common motif in nature, present in all kingdoms of life. The architectural simplicity of those linear repeat proteins is a major advantage for studying protein folding reaction. Compared to globular proteins repeat proteins lack long range interactions, allowing the dissection of energetics to different structural elements, which is required to construct energy landscapes.

The INK4 proteins are composed of ARs. Their four members ($p16^{INK4a}$, $p15^{INK4b}$, $p18^{INK4c}$, $p19^{INK4d}$) negatively regulate the mammalian cell cycle by specific inhibition of the two cyclin D-dependent kinases CDK4 and CDK6. Folding studies of $p19^{INK4d}$, consisting of five sequentially arranged ARs, revealed a kinetic intermediate during unfolding and refolding. A global analysis of CD- and fluorescence detected equilibrium folding transitions and the complex un- and refolding kinetics of $p19^{INK4d}$ confirmed a sequential folding pathway including a hyperfluorescent intermediate. This intermediate state populates only up to 15 % at equilibrium.

High resolution information on the intermediate state of $p19^{INK4d}$ was obtained by mimicking the earlier described phosphorylation sites of $p19^{INK4d}$ by glutamate mutations. A detailed analysis of NMR and fluorescence detected equilibrium and kinetic data of the $p19^{INK4d}$ S76E mutant confirmed, that the phosphorylation mimicking mutant corresponds to the earlier detected folding intermediate with the functional ARs unfolded whereas AR 3-5 remain folded. Ubiquitination of the double phosphorylation mimicking mutant $p19^{INK4d}$ S76E/S66E indicates a direct link between phosphorylation and ubiquitination.

Folding studies on naturally occuring AR proteins were until now focused only on eukaryotic proteins. To test the validity of a possible common mechanism of AR folding, a new AR protein in the evolutionary much older archaeal organism *Thermoplasma volcanium* was identified. The structure determined by X-ray crystallography confirmed that this archaeal AR protein (tANK) indeed folds into five sequentially arranged ARs with an additional helix at the N-terminus. Folding analysis of this protein revealed the same sequential three-state folding mechanism with the unusual fast equilibrium between the native and intermediate state as seen for $p19^{INK4d}$. GdmCl induced equilibrium unfolding transitions monitored by NMR gave high resolution information on the intermediate state of tANK since it could be populated to more than 90 percent under equilibrium conditions. Amide protons of AR 3-5 in the intermediate showed native chemical shifts whereas the N-terminal ARs are unfolded. Folding of AR proteins seems to follow a common principle: the most stable ARs fold first and provide a scaffold for the subsequent folding of the less stable but functional repeats.

The accumulation of folding intermediates raises the risk of protein misfolding and aggregation. Nature has evolved folding helper proteins to suppress this process and enhance productive folding. SlyD is one of them. The structure of *Thermus thermophilus* SlyD determined by X-ray crystallography revealed a two domain topology. By designing deletion constructs in combination with binding studies we could assign and map the binding interface of the chaperone function to the IF domain. The other domain, the FKBP domain, hosts the prolyl isomerase activity. Structures derived from two crystal forms differ in the relative orientation of both domains towards each other. They display different stabilities according to NMR detected H/D exchange and fluorescence equilibrium transitions. The two isolated domains are stable and functional in solution, but the presence of the IF domain increases the catalytic efficiency of the full length protein towards proline limited refolding of ribonuclease T1 100-fold. Therefore, we suggest that both domains work synergistically to assist folding of polypeptide chains. The combination of folding catalysis with a distal binding site for the folding protein chain is a common principle in nature.

Not all proteins or protein domains can fold on their own, because they require the presence of ligands, interaction partners, or a membrane environment. BAR (Bin/Amphiphysin/Rvs-homology) domains belong to a group of proteins which generate and sense membrane curvature. Their positively charged concave surfaces bind to the negatively charged membrane. Furthermore, N-BAR domains contain an N-terminal extension (helix-0) predicted to form an amphipathic helix only upon membrane binding. The structure and nano- to-picosecond dynamics of helix-0 of the human Bin1/Amphiphysin II BAR domain were determined in SDS and DPC micelles and confirmed the latter hypothesis. Molecular dynamic simulations of this 34 amino acid peptide revealed electrostatic and hydrophobic interactions with the detergent molecules, which induce helical structure formation from residues 8-10 towards the C-terminus. The orientation in the micelles was experimentally confirmed by backbone amide proton exchange. Both simulation and experiment indicate that the N-terminal region is disordered, and the peptide curves to adapt to the micelle shape. Deletion of helix-0 reduces tubulation of liposomes by the BAR domain, whereas the helix-0 peptide itself was fusogenic. These findings support models for membrane curving by BAR domains, where helix-0 increases the binding affinity to the membrane and enhances curvature generation.

Zusammenfassung

Proteine sind die funktionstragenden Moleküle in lebenden Zellen und deshalb auch wichtige Zielmoleküle in der Pharmazie und Biotechnologie. Um zu funktionieren, müssen sie eine definierte dreidimensionale Struktur annehmen. Die Abfolge der Aminosäuren, die Bausteine der Proteine, kodiert für diese bestimmte Konformation. Wie genau sich eine solche Polypeptidkette in ihre aktive Form faltet, ist eine essentielle Frage in der Strukturbiologie. Moderne, leistungsstarke Spektroskopiemethoden dienen dazu, diesen sogenannten Proteinfaltungsprozess aufzuklären.

Hauptfokus der vorliegenden Arbeit waren Faltungsstudien an Ankyrin-Repeat-(AR)-Proteinen. Der AR ist ein weitverbreites Strukturmotiv in der Natur, vertreten in Proteinen aller Lebensformen. Ihre stark vereinfachte Architektur bietet dabei einen grossen Vorteil bei der Untersuchung ihrer Faltungswege. Im Gegensatz zu anderen globulären Proteinen fehlen AR-Proteinen nämlich weitreichende Wechselwirkungen zur Stabilisierung. Dadurch können ermittelte Stabilitäten gewissen Strukturelementen zugeordnet werden, um so letztendlich Energielandschaften bestimmen zu können.

INK4-Proteine sind aus AR-Einheiten aufgebaut. Diese Gruppe zählt vier Mitglieder: $p16^{INK4a}$, $p15^{INK4b}$, $p18^{INK4c}$ und $p19^{INK4d}$. Sie regeln den Zellzyklus der Säugetiere durch spezifische Inhibition zweier Cyclin D abhängiger Kinasen, CDK4 und 6. Die umfangreiche biophysikalische Analyse des Faltungsmechansismuses von $p19^{INK4d}$, bestehend aus fünf AR-Einheiten, zeigte eine kinetische Zwischenstufe während der Ent- und Rückfaltung auf. CD- und Fluoreszenz-detektierte Gleichgewichtsfaltungsübergänge sowie komplexe zeitaufgelöste Faltungsanalysen von $p19^{INK4d}$ bestätigten einen sequentiellen Faltungsweg über diese hyperfluoreszierende Zwischenstufe (Intermediat). Dieses Intermediat populiert sich allerdings nur bis zu 15 % im Gleichgewichtszustand.

Detaillierte Informationen über die Zwischenstufe von $p19^{INK4d}$ wurden durch die Einführung einer Glutamatmutation erhalten, die eine bereits beschriebene Phosphorylierung in $p19^{INK4d}$ nachahmt. Tatsächlich bestätigte eine ausführliche Analyse der NMR- und Fluoreszenz-detektierten Gleichgewichtsmessungen und Kinetiken einer $p19^{INK4d}$ S76E-Mutante, dass die „Phosphorylierungsmutante" diesem zuvor entdeckten Faltungsintermediat enspricht, wobei die funktionellen ARs ungefaltet und AR 3 bis 5 gefaltet vorliegen. Nur dieser teilgefaltete Zustand kann in Zelllysaten ubiquitinyliert werden, was auf eine Verbindung von Phosphorylierung und Ubiquitinylierung hindeutet.

Bisher waren Faltungsstudien an natürlich vorkommenden AR-Proteinen nur auf eukaryotische Proteine beschränkt. Um die allgemeine Gültigkeit eines AR-Faltungsmechanismuses zu überprüfen wurde deshalb ein neues AR-Protein in dem evolutionär älteren Organismus *Thermplasma volcanium* identifiziert. Die Kristallstruktur zeigte, dass sich dieses archäische Protein (tANK) auch in fünf AR-Einheiten mit einer zusätzlichen Helix am N-Terminus faltet. Faltungsanalysen dieses Proteins offenbarten einen

Zusammenfassung

3-Zustandsfaltungsweg mit einem schnellen Gleichgewicht zwischen dem nativen und dem intermediären Zustand, wie bereits für p19^{INK4d} gezeigt. Strukturelle Informationen über das Intermediat von tANK resultieren aus NMR Messungen, da es sich bis zu 90 % unter Gleichgewichtsbedingungen populieren lässt. Amidprotonen der AR-Einheiten 3-5 des Intermediats zeigten native chemische Verschiebungen, wohingegen die N-terminalen AR-Einheiten ungefaltet sind. Die Faltung der AR-Proteine scheint einem allgemeinen Prinzip zu folgen: Die stabilsten AR-Einheiten falten sich zuerst und bieten dann den weniger stabilen aber funktionellen AR-Einheiten ein Gerüst für deren Faltung.

Das Auftreten von Faltungsintermediaten erhöht das Risiko der Proteinfehlfaltung und –aggregation. Um diesen Prozess zu unterdrücken und gleichzeitig die Produktivität der Proteinfaltung zu steigern, hat die Natur gewisse Helferproteine, die Chaperone, entwickelt. Hierzu zählt SlyD. Die Kristallstruktur von *Thermus thermophilus* SlyD zeigt eine 2-Domänentopologie. Mit Hilfe von Deletionskonstrukten und Bindungsstudien konnte die Chaperonfunktion der IF-Domäne zugewiesen und deren Binderegion charakterisiert werden. Die andere Domäne, die sogenannte FKBP-Domäne, birgt die Prolylisomeraseaktivität. Die Strukturinformationen stammen von zwei verschiedenen Kristallformen, die sich in der relativen Orientierung der beiden Domänen zueinander unterscheiden. Entsprechend den Ergebnissen des NMR detektierten H/D-Austausches und Fluoreszenzgleichgewichtsübergängen besitzen beide Domänen unterschiedliche Stabilitäten. Beide Domänen sind getrennt exprimiert strukturiert and funktionell, wobei die Anwesenheit der IF-Domäne die katalytische Aktivität des Volllängenproteins für die prolinlimitierte Faltung der Ribonuklease T1 100fach erhöht. Folglich arbeiten beide Domänen synergistisch. Diese Kombination aus Faltungskatalysator und einer distalen Bindungsstelle für die zufaltende Proteinkette ist ein bekanntes und verbreitetes Prinzip in der Natur.

Nicht alle Proteine oder Proteindomänen liegen ständig gefalten vor. Viele benötigen die Anwesenheit von Liganden, Interaktionspartnern oder eine Membranumgebung. BAR-(Bin/Amphiphysin/Rvs-Homologie-) Domänen gehören zu einer Gruppen von Proteinen, die gekrümmte Membranen erkennen bzw. deren Krümmung erzeugen können. Dazu binden ihre positiv geladenen konkaven Oberflächen an negativ geladene Membranen. N-BAR-Domänen besitzen eine N-terminale Erweiterung (Helix-0), von der man glaubt, dass sie bei Membranbindung eine amphipathische Helix ausbildet. Die NMR-Struktur und Nano- bis Picosekunden-Dynamiken der Helix-0 der humanen Bin1/Amphiphysin II BAR-Domäne in SDS- und DPC-Mizellen bestätigten oben genannte Hypothese. Molekulare Dynamiksimulationen dieses 34 Aminosäurenpeptids offenbarten elektrostatische und hydrophobe Wechselwirkungen mit den Detergenzmolekülen. Diese induzieren die Ausbildung der helicalen Struktur, angefangen bei den Resten 8-10 bis hin zum C-Terminus. Die Orientierung in den Mizellen wurde experimentell durch Rückgratamidprotonenaustausch bestätigt. Eine Deletion von Helix-0 der N-BAR-Domäne verringert ihre Tubulationsaktivität von Liposomen, während das Helix-0-Peptid selbst zur Fusion von Liposomen führte. Diese

Ergebnisse unterstützen Modelle, nach denen die Membrankrümmung durch diese BAR-Domänen induziert werden kann. Die Bildung der Helix-0 erhöht die Bindungsaffinität von BAR-Domänen an die Membrane und verstärkt die Krümmung.

1. Introduction
1.1 New View on Protein Folding

How does the amino acid sequence of a protein determine its three-dimensional structure? How does an inactive unfolded polypeptide chain fold up to its biologically active state?

An understanding of the molecular processes by which the one-dimensional sequence information of a polypeptide chain is converted into the three-dimensional structure of an active protein is an essential element in structural biology. Already 45 years ago Anfinsen and co-workers proposed that all the information necessary for a protein to achieve the native conformation (the so-called "protein folding reaction") in a given environment is encoded in its amino acid sequence [1]. But until now, the general mechanisms by which polypeptide chains fold into their defined three-dimensional structure are not well understood. Since Anfinsens pioneering experiments numerous protein folding studies have been carried out over the past few years and major insights into the nature of protein folding mechanisms are now emerging [2-6].

The native and biologically active state of a protein usually corresponds to the thermodynamically most stable structure under physiological conditions. The Levinthal paradox [7] dominated the ideas about the view of protein folding until recently. Levinthals concept was based on the random search problem. This means, that all conformations of the polypeptide chain (except the native state) have an equal probability, so that the native state can only be found by an unbiased random search. But the number of possible conformations for any polypeptide chain is so large, that a systematic search for a certain structure would take an astronomical length of time. Levinthal's solution to this folding problem was the prediction, that well-defined folding pathways exist to reach the native state. Meaning that protein folding is under "kinetic control" [7]. But in the last few years the picture changed. Approaches towards the protein folding problem now consider more the general characteristics of the energy surface of a polypeptide chain. This makes sense in a way, that the energy surface is one of the fundamental determinants of any reaction. Energy landscapes are therefore used to describe the search of an unfolded polypeptide guided along a funnel-like energy profile towards the native state [8-11] (Fig. 1).

Introduction

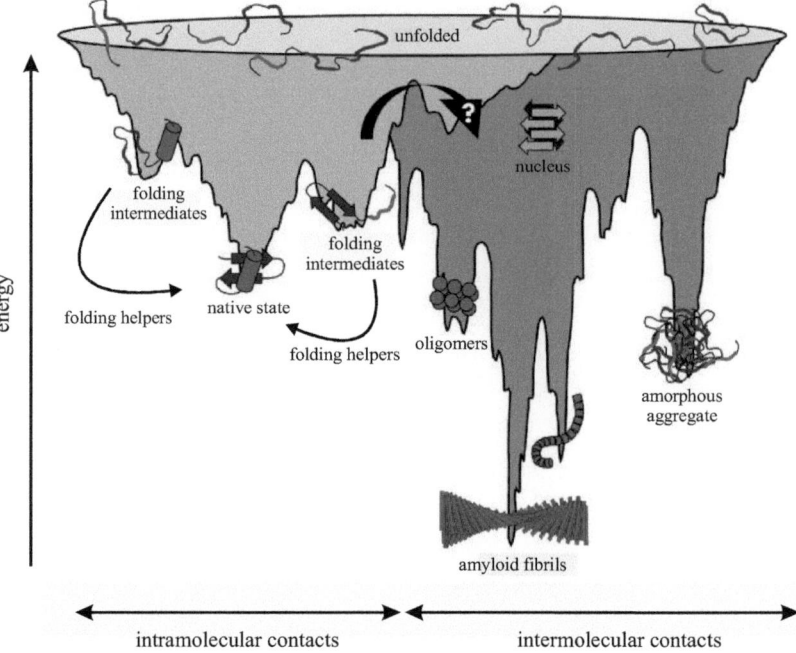

Fig. 1. Schematic energy landscape for protein folding and aggregation. The surface (grey) shows the large ensemble of unfolded states which "funnel" towards the native (light blue) or fibril state (dark blue) *via* partially folded intermediate states. The mechanistic details and further species to link partially folded states to fibrils are currently under investigation. High resolution models for intermediate states are barely available. So-called folding helpers are known to prevent partially folded states from aggregation and to speed up folding to the functional native state (after Jahn *et al.* [12]).

This idea refers to the new view on protein folding, which emphasizes the entire ensemble of protein conformations. The starting point of a protein folding reaction is no longer seen as one conformation of the denatured state, but rather as a very large collection of possible states. Consequently, a folding protein "funnels" to the global minimum state by various routes through the conformational space, which is considered as an energy landscape [13-15]. The polypeptide chain can find its energy minimum by a process of trial and error. Inherent fluctuations in the conformation of an unfolded or partially folded polypeptide chain facilitate contacts within the amino acid sequence, even between residues which are far separated in sequence. Because native-like contacts are more stable than non-native ones, they are more persistent, hence reducing the number of available conformations [10; 16; 17].

Since the energy landscape of a protein is encoded by its amino acid sequence, which has gone through evolution, only a reduced number of possible conformations has to be sampled by a given protein molecule during the transition from the unfolded to the native structure. Thus, natural selection has evolved proteins in such a way, that they are able to fold rapidly and efficiently by preventing misfolded states.

The surface of each folding funnel is unique and characteristic for a specific polypeptide sequence. It is determined by its thermodynamic and kinetic properties. Transitions between different states of protein folding can be investigated *in vitro* in detail by various spectroscopic techniques, ranging from optical methods to NMR spectroscopy [8; 18]. Changing pressure, temperature, or denaturant concentration (equation 1) affects the stability of a protein and allows to monitor folding/unfolding reactions of proteins. Therefore, rapid mixing devices (stopped-flow) and temperature/pressure jump machines were developed.

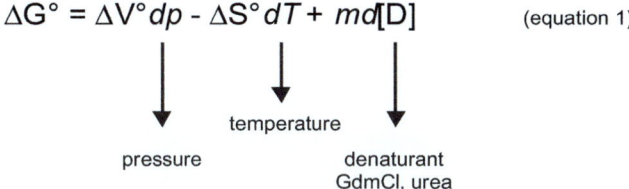

$$\Delta G° = \Delta V° dp - \Delta S° dT + m_d[D] \qquad \text{(equation 1)}$$

pressure   temperature   denaturant GdmCl, urea

Over the last 15 years, major advances in the field of protein folding resulted from studies on a series of small proteins (< 100 residues). They are easy to handle and can fold in the absence of complicating factors or intermediate states to the native structure in a cooperative two-state transition. Because their landscapes are relatively smooth these systems offered insights into the most basic steps of folding [19; 20].

Monitoring the effects of engineered mutations on unfolding and refolding kinetics of a specific protein allowed the analysis of the role of individual residues in the transition state of folding. These results suggested that the fundamental mechanisms of protein folding involve the interactions of a small number of residues to form a folding nucleus, around which the remaining amino acid residues rapidly condense [21-23].

The combination of experimental observations and computational simulations gave the first detailed picture of transition state ensembles. Despite a high degree of disorder, they show a similar overall topology as the native fold. But up to now, it is not clear how the sequence encodes these characteristics. However, the essential elements of a protein fold are mainly determined by the pattern of hydrophobic and polar residues. The latter favour the interaction of specific contacts as soon as the structure becomes more compact [3; 17]. Folding studies of proteins with more than 100 residues revealed in most cases the population of one or more intermediate states during the folding reaction, which might act as stepping stone to the native state [24].

There are controversial discussions about the significance of these intermediate species [25-28]. Do they assist folding to the native state by limiting the search process or do they act as kinetic traps and inhibit the folding process? Independent from this debate, structural properties of intermediate states provide important insights into the folding of larger molecules. Even more important, partially folded states are intrinsically prone to aggregation, which ultimately leads to protein misfolding (Fig. 1). A generally accepted hypothesis for protein aggregation is the exposure of hydrophobic regions of partially folded states populated during folding reaction or caused by local fluctuations of the native state. Aggregates can be either ordered with fibril morphology or amorphous, resulting in inclusion bodies [29].

The mechanisms for these specific aggregation events have drawn intense interest by the protein folding community [30; 31]. Now, folding studies expand their impact from key fundamental principles to a direct understanding of several human diseases. Amyloid fibrils are found as deposits of insoluble aggregates in Alzheimer's, Parkinson's, Creutzfeld-Jacob disease, or type II diabetes [32].

NMR and X-ray crystallography give a detailed picture of the structural and dynamic properties of proteins in their native state. However, high resolution structural information on intermediate or misfolded states is more difficult to achieve. Their rapid interconversion, low population under equilibrium conditions, and high aggregation tendency renders them difficult to analyze by classical approaches. But using modern NMR methods combined with dynamic simulation and trapping intermediates by mutations, starts to shed light on the structural properties of these partially folded state ensembles on the folding energy landscape [33-35]. A combinatorial approach (Fig. 2), including different techniques from various scientific fields, will help to gain a detailed understanding of folding intermediates on their route to the native state.

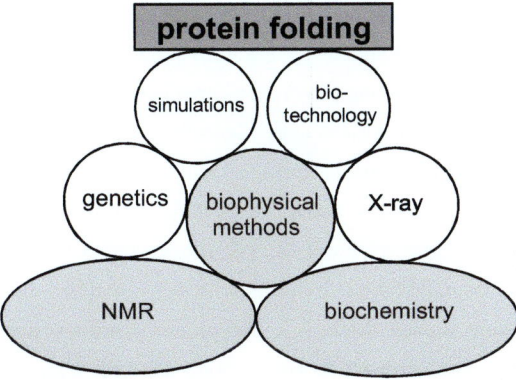

Fig. 2. The combination of experimental and computational techniques from different scientific fields will be necessary to improve the understanding of protein folding.

1.2 Protein Folding in the Cell - Chaperones

The basic understanding of protein folding has been achieved so far from *in vitro* and *in silico* folding studies. Although it has been shown that *in vitro* folding is a valid model for understanding the folding process of a nascent polypeptide chain *in vivo*, there are still several differences between the cellular environment and the test tube.

The interior of a cell is densely packed with macromolecules, like proteins, DNA, RNA, and polysaccharides [36; 37]. Some proteins can only fold in specific compartments, such as mitochondria or the endoplasmatic reticulum after trafficking and translocation through membranes [38; 39]. Although many details of the folding process in a cell depend on a particular environment, the fundamental principles of folding are universal.

As a result of the molecular crowding effect in a cell, partially folded proteins unavoidably expose regions which are buried in the native state to the surroundings. This may lead to unfavourable interactions, accumulation, and finally misfolding or aggregation. But living systems have evolved folding helper proteins, so-called chaperones, to prevent or correct such behaviour. Chaperones are present in all types of cells and compartments. They interact with nascent polypeptide chains or assist folding in later folding processes [39-42]. Usually a network of different chaperones is necessary to ensure efficient and complete folding. Functional details of chaperones mainly result from *in vitro* studies. Up to date, more than 20 chaperone families have been identified, but most of them show no or just little substrate specificity [43; 44].

To prevent aggregation or premature folding they bind transiently to hydrophobic regions of unfolded or partially folded polypeptide chains ("holdases") [45]. It is also well known, that molecular chaperones are able not only to protect proteins as they fold, but also to rescue misfolded or already aggregated proteins giving them another chance to fold correctly. This active participation during the folding process consumes energy [38; 39]. Therefore, most but not all molecular chaperones require ATP as energy source. However, chaperones themselves do not increase the rate but the efficiency of folding by reducing the risk of aggregation.

Beside chaperones, there are several classes of folding catalysts, which speed up potential slow steps in folding and therefore lower the risk of aggregation [46]. Among these the most prominent ones are peptidyl-prolyl isomerases, which catalyze the rate of *cis/trans* isomerization of proline peptide bonds [47-49], and protein disulfide isomerase [50; 51], which increase the formation rate of disulfide bonds. Some of these enzymes provide catalytic activity as well as chaperone function, localized in an additional domain [52-56]. The prolyl isomerase SlyD shows this dual topology and their structural and functional characterization is part of this thesis. The chaperone and enzymatic domains are suggested to work synergistically to assist folding of polypeptide chains. But the detailed interplay is not well understood.

1.3 Repeat Proteins

Proteins containing repeating amino acid sequences have drawn great attention in the last few years. Based on recent developments in sequencing technology, complete genomes of numerous organisms became available. They revealed that short, tandem repeating motifs are common in many proteins throughout all kingdoms of life [57; 58]. Nearly 20 percent of all proteins encoded by the human genome contain repeating units of 20-40 amino acid. These building blocks stack onto each other forming a modular, elongated architecture with a specific protein binding interface, different compared to globular proteins [59-62]. Internal duplication, insertion, deletion, and recombination are the simplest explanation for the existence and development of these repeat proteins.

The modular architecture may be the key to their evolutionary success. Simple multiplication of existing genetic material enables an organism to evolve protein sequences faster and thus to rapidly adapt to new environments. Therefore, it is not surprising that repeat proteins are most common in eukaryotes, due to their increased complexity of cellular functions [58; 60].

In contrast to SH2 or SH3 domains, which are also known to mediate protein-protein interactions, these modular repeat proteins do not recognize a specific sequence but rather determine their specifity for partner proteins by variations in their adaptive surface residues. Sequence analysis revealed consensus sequences for different groups of repeat proteins: While conserved amino acids form the repeating structural building block, non-conserved amino acids located mainly on the surface generate a high variability of protein binding surfaces [63; 64].

In combination with selection methods such as ribosome or phage display, it recently became possible to construct synthetic repeat proteins with high specificity for target proteins. These designed repeat proteins are thermodynamically more stable than their natural counterparts, and advantageous compared to antibodies, because they lack disulfide bridges and can easily be produced in *E. coli* strains [65-70].

1.4 Structure and Classification of Repeat Proteins

Repeating modules of repeat proteins contain secondary structure elements that can fold in a variety of topologies (Table 1). The linear assembly of repeats results in a simple scaffold, which is dominated by mainly hydrophobic short range interactions within or with adjacent repeats. In general, sequentially distant residues (residues of non-adjacent repeats) do not interact with each other. Numerous stabilizing long range interactions, causing complex topologies in globular proteins, are absent in repeat proteins. The lack of long range interactions in combination with this simple architecture makes repeat proteins an exciting and easy-to-handle system to study protein folding, stability, function, and design [71; 72]. Table 1 shows a selection of commonly occurring repeat proteins classified according to their architecture. β-propeller repeats are omitted because the radial arrangement of their repeats leads to a propeller like architecture, which is more similar to globular structures.

Introduction

Table 1. Structure of various repeat proteins. Architecture of a single repeat is described and displayed: α-helices are red, β-strands are yellow, polyproline II structure is green and turns are blue. Examples for each repeat class show the linear array of the same repeat, with adjacent repeats coloured from red to purple (ankyrin repeats of the Notch receptor, 1O8T.pdb; heat repeats, 1UPK.pdb; leucine-rich repeats, 1H62.pdb; hexapeptide repeats, 1J2Z.pdb; tetratricopetide repeat, 2F07.pdb; after Kloss et al. [72]).

Repeat type	Architecture	Single Repeat	Example of Structure
Ankyrin repeat	33-residue motif forming a helix-loop-helix-α-turn motif, which is L-shaped in cross-section		
Heat repeat	37-47 residue motif, comprising a pair of antiparallel helices, with a characteristic kink of the first helix		
Leucine-rich repeat	20-29-residue motif forming a β-strand-loop-helix structure		
Hexapeptide repeat	hexapeptide motif, comprising a β-strand and loop, forming a continuous β-helix		
Tetratricopetide repeat	34-residue motif with a pair of antiparallel helices		

1.5 Ankyrin Repeats (ARs)

The ankyrin repeat (AR) is one of the most common motifs in nature, present in all kingdoms of life, including bacteria, archaea, and eukarya, as well as in viral genomes [60]. The SMART database compasses nearly 17,000 AR containing sequences with the majority (> 90 percent) found in eukaryotes (date: 12th March, 2008). In contrast, just eight archaeal protein sequences are assigned with a putative AR fold [61]. The AR motif shows a canonical helix-loop-helix-β-hairpin/loop topology. The helices are arranged in an antiparallel manner, connected by a tight turn and followed by a β-hairpin. The helices of one repeat pack against the helices of adjacent repeats, while β-hairpin/loop regions can form a continuous β-sheet. Hydrophobic interactions between neighbouring helices in addition to a hydrogen bonding network are characteristic for AR proteins. Their overall shape exhibits a slight curvature, caused by the differences in helix length and the interrepeat packing interactions between the two helices [73; 74].

The AR motif was first identified in the yeast cell cycle regulator Swi6/Cdc10 and the signalling protein Notch from *Drosophila* [75]. But its name is presumably derived from the cytoskeletal protein ankyrin, which contains 24 copies of this repeat [76]. AR proteins can exist as single proteins or in combination with other domains in multidomain proteins. Up to 33 ARs are found in a single protein, but the majority of proteins contain less than six repeats [57]. One isolated AR can not adopt a folded structure due to its intrinsic instability. Therefore, the minimum folding unit of isolated AR proteins was determined to two [77].

AR proteins participate in a wide range of cellular functions, including cell-cell signalling, transcription, and cell-cycle regulation or various transport phenomena [57]. Their typical function is the mediation of protein-protein interactions. One of the biologically most important and structurally best characterized group is the family of INK4 tumour suppressors. Its four members (p15^{INK4b}, p16^{INK4a}, p18^{INK4c}, p19^{INK4d}) are all AR proteins and negatively regulate the cell cycle by inhibiting cyclin dependent kinases (CDKs) 4 and 6, which trigger the progress of the cell cycle from the G1 to the S-phase [78; 79]. Mutations found in the INK4 family are directly linked to cancer [80].

1.6 BAR Domains and Membrane Curvature

The assembly of lipids into bilayers to generate membranes is fundamental for cellular structure and compartmentalization [81; 82]. For plasma membranes or membranous organelles it is often necessary to change the membrane shape [83]. This involves the formation of high curvature microdomains to generate tubules or vesicles. Nature has evolved curvature inducing and stabilizing protein modules, that can shift the equilibrium between different membrane curvature states. The BAR (Bin, Amphiphysin, Rvs) domain superfamily (Fig. 3) has been identified as important actor in membrane remodeling processes througout eukarya [84-88]. Members are recruted from the cytoplasma to trigger the formation of plasma membrane extensions, invaginations, and transport intermediates like endocytic vesicles. Most information about structure-function relationships of the BAR superfamily results from crystallographic studies, showing its members are elongated, antiparallel dimers of α-helical coiled coils [87; 89]. They differ in their overall degree of curvature (compare F-BAR and classical BAR domains) suggesting, that they sense and bind to different curved membranes. Despite these differences, the surface of all BAR domains shows clusters of positively charged patches, which are positioned to interact with negatively charged phospholipid headgroups of the membrane.

Different mechanisms for BAR domain induced curvature have been discussed [90]. A high surface densitiy of BAR domains seems to be required to initiate membrane tubulation. This implies a cooperativity effect between certain BAR molecules, either through protein-protein interactions, membrane-mediated interactions, or domain ordering caused by protein crowding [91-93]. Furthermore, the group of N-BAR domains contains an N-terminal extension which is predicted to form an amphipathic helix upon membrane binding but is unresolved in all crystal structures. The insertion of a helix into one leaflet of the bilayer induces local bending and is proposed to significantly contribute to membrane curvature [88; 94; 95].

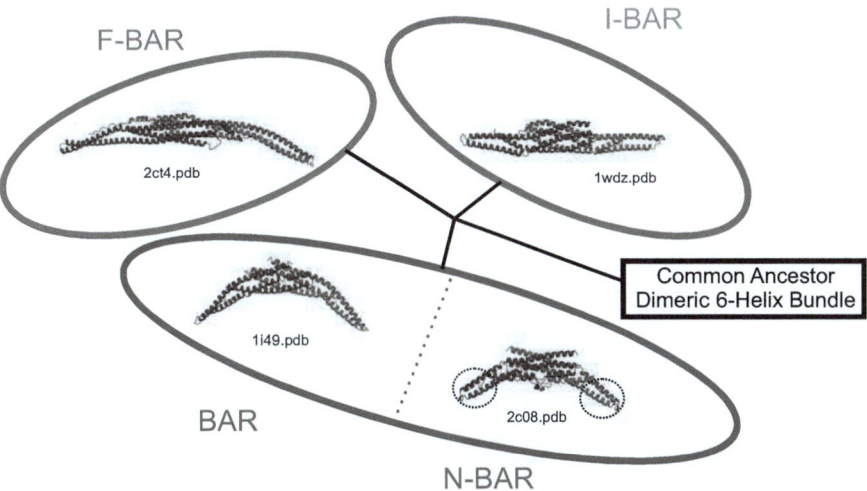

Fig. 3. The BAR domain superfamily. Representantive members of the BAR superfamily are given with corresponding pdb-code. F-BAR for FCH and BAR, I-BAR for "inverse" BAR and N-BAR for the conjunction of N-terminal membrane penetrating amphipathic helices with a BAR domain, but the amphipathic helix is not resolved in the crystal structure (expected in the dashed circle).

1.7 Purpose of This Thesis

Advances in computational and experimental studies have greatly improved the understanding of protein folding in the last 25 years. Characteristic for folded globular proteins is the close proximity of distant segments in the polypeptide chain. Therefore, it is not surprising that topologies based on long range interactions promote cooperativity in protein folding and prevent independent structural fragments from folding. This makes the mapping of energy landscapes difficult. Repeat proteins bypass this problem, because their elongated structures lack long range interactions.

The modular nature of AR proteins raises a lot of interesting questions from a protein folding point of view. Do repeat proteins fold and unfold in a cooperative or non-cooperative manner? Or does this repeat architecture support a multi-state folding pathway with intermediate states consisting of some repeats folded and others unfolded? And if so, what are the characteristics of this intermediate states? Are they important for folding or do they act as kinetic traps? Contrary to this speculations, initial folding studies on naturally occurring AR proteins displayed highly cooperative equilibrium unfolding transitions without any detectable partially folded intermediate states under equilibrium conditions.

The human AR protein p19^{INK4d} from the INK4 family was choosen as a model protein for a rigorous analysis of the folding mechanism of a naturally occuring AR protein to adress above mentioned questions. Since p19^{INK4d} is devoid of any fluorophores, it was necessary to introduce tyrosine or tryptophan residues as fluorescent reporter groups in certain repeat segment without changing function and stability of the protein. By using a global analysis approach, all biophysical parameter for a folding model of p19^{INK4d} should be extracted. To correlate results of *in vitro* experiments with *in vivo* function the obtained p19^{INK4d} folding mechanism should then be discussed and analyzed in context of a functional and cellular background.

Goal of the second part of this work, was to identify and study AR proteins of much older organisms like archea. The database predicts only a few archaeal AR proteins, which makes it highly interesting to compare protein folding data of proteins, which are very similar in terms of structure and function, but far distant in evolution.

The folding reaction is essential for each protein to reach its native state and to become active. Several reasons, which had already been discussed, can limit the folding reaction leading to misfolding and aggregation (see chapter 1.2). Therefore, in the third part of this thesis, we focused on enzymes with chaperone function, which speed up protein folding reactions and prevent aggregation. The prolyl *cis/trans* isomerase SlyD was choosen as a target protein since structural information was still lacking. SlyD is found in different organisms and is involved in various cellular processes. Sequence analysis proposed a simple architecture based on two domains hosting different functions. This principle is widely found in nature and attracts SlyD as a model protein for a two domain protein study. To understand these systems in more detail a combinatorial approach using different methods, ranging from biochemistry, X-ray, NMR, and SAXS should be applied.

Biological membranes and membrane proteins have drawn strong attention in the biochemistry field within the last years. In the last part of this thesis, recently identified BAR domains, which are able to sense and curve membranes, were the research focus. They play fundamental roles in membrane fusion, budding, or tubulation. N-BAR domains contain an N-terminal extension, which is suggested to fold from a random coil structure into a helical conformation upon membrane binding. Due to its intrinsic disorder in solution, this extension is unresolved in all crystal structures, but essential for function. The goal therefore was to study the structural and dynamic properties of the proposed amphipathic helix in lipids and detergent micelles and unravel their functional importance.

2. Summary and Discussion

2.1 Protein Production

To study structure and folding reactions of biological macromolecules, large amounts of highly pure protein are necessary. In the early days, proteins were obtained from natural sources, rich in particular proteins. Isolating proteins directly from tissues is still a common practice in biochemistry [96], but has great limitations because it requires a high abundance of a specific protein in certain cell types or tissues.

The development of recombinant DNA technology was a milestone for protein production [97]. Suddenly, it became possible to introduce a specific gene into a fast growing organism like *E. coli*, in order to produce the protein of interest in sufficient amounts. Furthermore, now mutations could be easily introduced and their influence on structure, stability, and function studied.

Biotechnological industry takes advantage of recombinant expressed proteins, because besides being cost effective, they are chemically identical to their naturally produced counterpart, lowering the risk of immune response [98]. Over the past few years, a variety of expression systems, including multiple choices of strains, expression vectors, promoters, and different purification tags were developed for protein production and purification. Nevertheless, the overexpression of recombinant proteins remains challenging. In most cases the production of soluble, active protein is desired, but fails due to toxicity or folding problems in the cell. Latter may lead to the formation of inclusion bodies or degradation of the protein by the host organism [99]. Eukaryotic proteins are considerably more challenging. They often require a complex chaperone system or posttranslational modifications like glycosylation for soluble expression, accomplished by enzymes, absent in a prokaryotic host as *E. coli*. To overcome these problems and to increase the soluble protein yield, numerous strategies are employed. Screening of multiple parameters as expression temperature, media and additives, strains, inducer concentration, expression time, and point of induction has been shown to be beneficial for the soluble and active overexpression of a large set of proteins [100-102].

In this work a variety of proteins and their mutants from different organisms were overexpressed in high amounts and purified to homogeneity for structural and folding studies. For that purpose, the following strategy (Fig. 4) was developed and applied for all proteins for an effective, fast, and reproducible protein production.

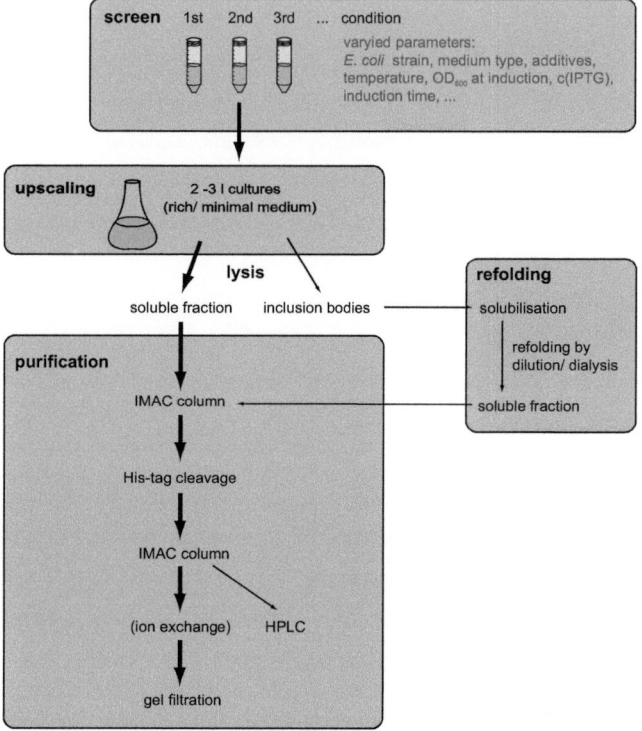

Fig. 4. Flow schema for overexpression and purification of recombinant expressed proteins. First, conditions were optimized for soluble expression in small scale. Proteins resulting in inclusion bodies were subjected to different refolding conditions and in successful cases purified to homogeneity.

Soluble protein expression was optimized in small scale by varying the earlier mentioned parameters. All overexpressed proteins contained a His-tag to facilitate affinity purification *via* an IMAC column, followed by tag-cleavage using thrombin (except for SlyD) and gel filtration. Proteins with insufficient purity (e.g. p19^{INK4d} mutants) revealed by SDS-PAGE were subjected to a further purification step using ion-exchange chromatography. The final polishing step for recombinantly expressed peptides (e.g. BAR peptide) was performed by RP-HPLC prior to lyophilisation. Nevertheless, certain proteins and strongly destabilized mutants (e.g. CDK6, p19^{INK4d} mutants containing a S76E mutation) could not be overexpressed in soluble form, because they accumulated as inclusion bodies in the cytosol of the host strain. Therefore, a refolding protocol was established for each of those target proteins.

Inclusion bodies were washed and solubilised in denaturants e.g., urea or GdmCl, and refolded by dilution or dialysis. Soluble material was further subjected to the purification work flow (Fig. 4) Correct folding state was checked by the gel filtration profile, CD spectroscopy, 1D and 2D NMR, as well as by activity assays.

Soluble overexpression and purification of CDK 6 was a major challenge. Although trying different constructs, covalent fusions or coexpression of chaperones, protein expression in *E. coli* always resulted in inclusion bodies. More than 50 refolding conditions were sampled following analytical gel filtration, without any success. Eventually, the complex baculovirus/insect cell system led to soluble overexpression of CDK6 (collaboration with Prof. ED Laue, Cambridge) [103]. Protein yields were comparatively low but sufficient for binding studies with p19.

2.2 Folding Mechanism of CDK Inhibitor p19^{INK4d}

p19^{INK4d} (p19) consists of five sequentially arranged ankyrin repeats (AR) and controls the human cell cycle by inhibiting CDK 4 and 6. Inhibition of CDK4 and 6 is mainly mediated by AR 1 and 2 [78; 103; 104]. We used p19 as model protein to study the folding mechanism of AR proteins by equilibrium and kinetic experiments (Fig. 5).

Wild type p19 is devoid of tyrosine and tryptophan residues, and thus lacks sensitive fluorescent probes for the folding analysis *via* fluorescence spectroscopy. This allowed a site specific introduction of fluorescent amino acid in certain ARs. Several positions were tested and analyzed according to stability, fluorescence change between native and unfolded molecules, and function (Table 2). Tryptophan 86 turned out to be the best reporter, because it did neither affect function nor stability compared to wild type p19. Furthermore, it provided the most sensitive probe for kinetic folding studies. Expectations that fluorescent probes in certain ARs could monitor local folding events (e.g. folding of single repeats) were not supported. We could rather confirm that a single reporter can globally probe folding reactions. In addition, truncated variants of p19 could be produced. These constructs were lacking either the fifth (p19 AR1-4) or the first two ARs (p19 AR3-5). They were globally folded as judged by CD and NMR spectroscopy. As expected, p19 AR1-4 still bound to CDK6, while the truncation of the first two ARs (p19 AR3-5) abolished CDK6 binding (Fig. 5).

Summary and Discussion

Table 2. Purified p19^{INK4d} mutants used for folding studies. "AR" column shows repeat number, where mutation is localized.

Mutation	AR
G19W	I
L32W/Y	I
H34W	I
F41W/Y	I/II
F51W/Y	II
T75W	III
F86W/Y	III
H96W/Y	III
T106W	III/IV
H119W	IV
F125W/Y	IV
Q148W/Y	V
I157W	V
H96W (p19 AR1-4)	III
F86W (p19 AR3-5)	III
F41W/T106C	II, IV
F86W/S76E	III, II
F86W/S66E	III, II
F86W/S76E/S66E	III,II,II
F86W/S76A	III, II
H96W/S76E	III, II

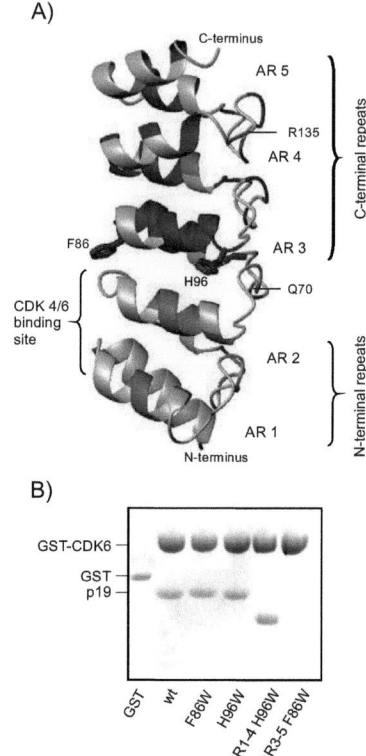

Fig. 5. (A) Crystal structure of p19^{INK4d} (1bd8.pdb from the Protein Data Base). Five ARs (AR 1-5), each comprising a loop, a β-turn, and two sequential α-helices form the elongated structure, where Phe86 and His96 are indicated by a stick illustration of the side chains. Residues with highly protected backbone amide protons against solvent exchange (P > 12000) using NMR H/D exchange are indicated in blue and less protected backbone amides (P < 12000) in red. Q70 denotes the first residue of truncated p19 AR3-5 containing C-terminal ankyrin repeat 3, 4, and 5 of the wild type protein and R135 denotes the last residue of truncated p19 AR1-4. The CDK4/6 binding site is mainly formed by the N-terminal ankyrin repeats 1 and 2. SDS-PAGE analysis of the pull-down assay of wild type p19^{INK4d} and different variants. (B) Immobilized GST-CDK6 on glutathione sepharose bound wild type p19, p19 F86W, p19 H96W, and p19 AR1-4, whereas p19 AR3-5 did not bind.

Urea and temperature induced equilibrium transitions monitored either by CD or fluorescence spectroscopy were analyzed according to a two-state model for all variants (except mutants carrying the S76E mutation; see chapter 2.3) without evidence for intermediate states. This is in agreement with earlier works for natural occurring AR proteins, where partially folded intermediate states were not detectable in equilibrium unfolding transitions [105-107].

Summary and Discussion

The cooperativity of the folding transition is comparable to globular proteins of similar size. This indicates that a certain type of coupling mechanism exists among different repeats. The stability of ≈ 28 kJ/mol and an unfolding midpoint close to 3 M urea makes p19 to the most stable member of the INK4 family. Compared to the four AR comprising tumour suppressor p16, the stability of p19 is 2.5 fold increased. A possible reason might be the additional stabilizing interactions of a fifth AR in p19. In line with this hypothesis, the deletion of AR5 in p19 strongly reduced the stability to a similar ΔG_{NU}-value as found for p16. A fragment comprising AR 3-5 of p19 unfolds and refolds reversible, but displays a strongly reduced stability compared to the full length protein.

Single and double mixing stopped-flow fluorescence spectroscopy experiments of p19 F86W gave detailed insights into the folding mechanism of p19 (Fig. 6-8). One of its characteristics is the biphasic unfolding process with a hyperfluorescent intermediate state (Fig. 6A). Upon unfolding, the fluorescence of Trp86 strongly increased, reached a maximum, and finally decreased to a value which is slightly lower than the fluorescence of the native state. This clearly shows that a hyperfluoresent intermediate state becomes populated during the unfolding reaction. The refolding reaction of unfolded p19 F86W molecules is also biphasic, but for both phases the fluorescence signal increased (Fig. 6B).

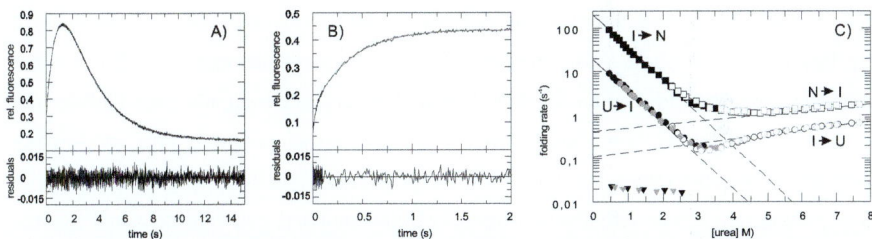

Fig. 6. (A,B) Single mixing unfolding and refolding kinetics of p19 F86W monitored by stopped-flow fluorescence. (A) Unfolding was initiated by a rapid change from 0 M to 6 M urea. (B) Refolding was initiated by rapid dilution from 6 M urea to 0.9 M urea. Data can be best described by a double exponential function. Deviations of the fits from the experimental data are given as residuals below each trace. (C) Urea dependence of apparent folding rates of p19 F86W. Filled symbols indicate refolding experiments, open symbols indicate unfolding experiments. Folding rates of the fast phases from stopped-flow experiments are represented by squares and slow phases by circles. Very slow refolding rates from manual mixing experiments are depicted as triangles. Grey symbols represent folding rates determined by stopped-flow CD. The solid line in (C) represents the result from a global analysis of kinetic and equilibrium data. Dotted lines represent the urea dependence of the intrinsic rate constants for the indicated reaction. The errors for the rate constants are smaller than the symbol size.

Folding rates derived from far UV-CD stopped-flow spectroscopy experiments agreed well with fluorescence data, except that the second fast refolding phase is not detectable. Kinetic data derived from other Trp mutants showed a similar behaviour with two unfolding and one refolding rate (data not shown).

Kinetics of the truncated variant p19 AR3-5, lacking the two functional N-terminal repeats, are monophasic under all conditions. In addition, p19 and the deletion variant showed a very slow refolding phase which is limited by prolyl isomerization.

To understand and assign the complex folding kinetics of p19, double mixing experiments were carried out, starting from either the native (`U-assay`) or the unfolded state (`N-assay`) (Fig. 7). The `N-assay` provides information on the rate-limiting step(s) of folding by monitoring the population of intermediate and native species during the refolding reaction. Therefore, the refolding reaction was interrupted at various time points and native as well as intermediate molecules subsequently transferred to unfolding conditions. With increasing refolding time, more molecules became native and the unfolding amplitude increased. Since unfolding of p19 is biphasic, amplitude progression of both reactions with increasing refolding time could be analyzed, but did not reveal significant different behaviour.

A double exponential fit to the amplitude dependence upon refolding time revealed similar rate constants for both data sets (Fig. 7C). The rate constants are in good agreement with the slow and the slowest rate constants derived from single mixing experiments. Together with the lack of the fast phase during this experiment this clearly demonstrated that the slow refolding rate is rate-limiting for the formation of N molecules during refolding. Furthermore, it showed, that the slowest phase, caused by prolyl isomerization, is also rate-limiting. Together with earlier reported real-time NMR experiments it can be concluded, that this very slow folding step is caused by the heterogeneity of the unfolded state ensemble, and a certain fraction of polypeptide chains has to isomerize prior to folding.

Summary and Discussion

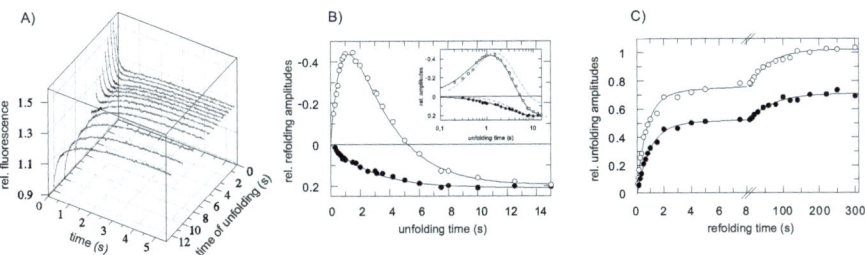

Fig. 7. (A,B) Double mixing protein folding experiments of p19 F86W to monitor species during unfolding ('U-assay'). Unfolding was initiated by a rapid change from 0 M to 4.5 M urea. After various times of unfolding, the subsequent refolding reaction under fluorescence detection was started by a second fast change to 1.5 M urea. (A) Double exponential equations (red lines) were fitted to the fluorescence intensities (solid black lines). (B) The amplitudes from these fits are shown with open symbols for the fast folding phase and for the slow phase with closed symbols at different times of unfolding. A fit of a double exponential function to the open symbols gave rate constants of 1.22 ± 0.08 s^{-1} and 0.38 ± 0.02 s^{-1}, the fit of a single exponential function to the closed symbols gave 0.35 ± 0.02 s^{-1} (continuous lines). The inset shows the same data plotted on a logarithmic time scale. The dotted line represents a simulation of the 'U assay' from the intrinsic unfolding and refolding rates derived from the global fit. (C) Double-mixing experiments to monitor intermediate and native species during refolding ('N-assay'). Amplitudes of the fast unfolding phase of p19 F86W at 6 M urea after variable refolding times at 1.5 M urea are depicted with closed symbols and amplitudes of the slow unfolding phase by open symbols. A fit of a double exponential function to the amplitude dependence upon refolding time (solid lines) revealed for both data sets rate constants of 0.68 ± 0.02 s^{-1} and 0.018 ± 0.002 s^{-1}.

To correlate the observed hyperfluorescent intermediate state during unfolding with refolding kinetics, a second set of double mixing experiments was carried out. Native p19 F86W was first diluted from 0 to 4.5 M urea to initiate folding. After different time periods, unfolding was stopped and refolding started by dilution to 1.5 M urea (Fig. 7A,B). The resulting refolding kinetics showed that the fast refolding kinetic (observed also in single jump kinetics) is directly linked to the I → N reaction, whereas U → I is slow. Taken these findings together, the folding mechanism drawn in scheme 1 can be concluded:

$$U_{cis} \rightleftarrows U_{trans} \overset{\text{rate limiting}}{\rightleftarrows} I \overset{\text{fast}}{\rightleftarrows} N \qquad \text{(scheme 1)}$$

The formation of the intermediate state is rate-limiting and slow in the folding pathway of p19. As soon as the intermediate state is reached, the molecules fold to completion in a fast manner. The slow isomerization reaction is caused by the heterogeneity in the unfolded state and ~ 20 percent of the molecules have to isomerize prior to folding.

Based on this mechanism, the fast refolding phase (I → N) should not be visible in the single mixing kinetics. Indeed, this folding phase is absent in most mutants (except mutants with hyperfluorescent I) and when folding is monitored by CD. However, the high fluorescence of the intermediate state compared to the native and unfolded state causes a situation, where the fast refolding phase becomes visible during refolding, despite its rather small amplitude. This explanation was verified by simulating the refolding kinetics based on the intrinsic rate constants (data not shown). The folding mechanism of the deletion mutant was simplified, since intermediates are not populated during folding and unfolding. But similar to the full length protein a certain fraction of molecules showed non native prolyl isomers in the unfolded state which have to isomerize prior folding.

A global analysis including kinetic and equilibrium data resulted in the intrinsic rate constants and cooperativity values of p19 folding (Table 3). This allowed the calculation of native, intermediate, and unfolded populations under equilibrium conditions at a given urea concentration. Interestingly, according to this data, the intermediate state can be populated up to 15 percent under equilibrium conditions (Fig. 8C, dotted line). This is presumably too low to produce a pronounced deviation from a two state equilibrium unfolding curve. To verify these findings we developed a sensitive kinetic assay (`I-assay`) (Fig. 8). P19 F86W was preincubated at various urea concentrations prior to unfolding into the same final urea concentrations. Since all native molecules unfold *via* the hyperfluorescent intermediate, the ratio of both unfolding amplitudes should be constant, if only native and unfolded molecules would be present under equilibrium conditions. But this is clearly not the case, because the amplitude for the fast unfolding reaction (N → I) declined at lower urea concentrations compared to the amplitude for the slow unfolding reaction (I → U).

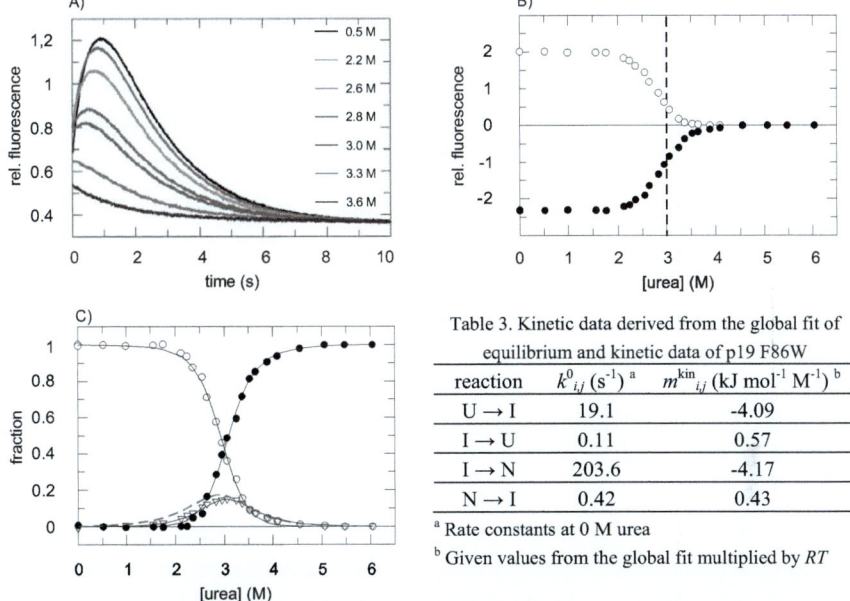

Fig. 8. (A-C) Assay for intermediate state of p19. Unfolding of p19 F86W was monitored at 6.6 M urea after equilibration of the protein at urea concentrations between 0 M and 6 M urea. (A) Unfolding kinetics of p19 F86W incubated between 0.5 M and 3.6 M. (B) Amplitudes of the fast phase (1.48 ±0.03 s^{-1}; open symbols) and the slow phase (0.53 ± 0.02 s^{-1}, closed symbols) of unfolding as a function of the urea concentration used for equilibration. Double exponential functions were fitted to the measured kinetics. The dotted line represents 3 M urea close to the maximum population of I. (C) Calculated equilibrium populations of the native (open symbols), intermediate (triangles), and unfolded state (closed symbols). The dotted line represents the expected population of I calculated from the urea dependence of the intrinsic un- and refolding rate constants derived from the global fit (Table 3).

Table 3. Kinetic data derived from the global fit of equilibrium and kinetic data of p19 F86W

reaction	$k^0_{i,j}$ (s^{-1}) [a]	$m^{kin}_{i,j}$ (kJ mol^{-1} M^{-1}) [b]
U → I	19.1	-4.09
I → U	0.11	0.57
I → N	203.6	-4.17
N → I	0.42	0.43

[a] Rate constants at 0 M urea
[b] Given values from the global fit multiplied by RT

The amplitude values of this kinetic unfolding assay allowed a direct calculation of the native, intermediate, and unfolded fraction at a given urea concentration. Experimental data agreed well with calculated populations derived from the global fit and clearly showed that the intermediate state is populated under equilibrium conditions although just to a low extent. The kinetic analysis gave information on the rates of interconversion between N, I, and U and its stability, but not on the structural properties of I.

A comparison of biophysical data of the full length protein and the truncated variant p19 AR3-5 proposed that AR 1 and 2 are still unfolded or incompletely folded in I. To validate this speculation, the local stability of p19 was measured by NMR H/D exchange. Resulting protection factors revealed remarkable differences in the stabilities of individual repeats. AR 3 and 4 show the highest stability, whereas amide protons of AR 5 are, on average, 10-fold and those in AR 1 and 2 ~ 100-fold less protected. Calculated stabilities for AR 1 and 2 based on the protection factor agree well with the ΔG_{NI}-value derived from kinetics. These data further support a model, that AR 3-5 fold first and provide a scaffold for the less stable but functionally important ARs 1 and 2 (for details see subproject A).

2.3 P19^{INK4d} Between Native and Partially Folded State

High resolution information on structural properties of folding intermediates is limited in literature. Reasons therefore are the low population under equilibrium conditions and their high tendency for aggregation. Mutational analysis is a widely used technique to trap intermediate states at equilibrium [35; 108]. However, often it is not clear, whether the structure of the mutated protein really reflects the intermediate state of the wild type protein.
There is evidence that mutations in AR proteins can change their folding behaviour, limiting the validation of the wild type mechanism [109; 110]. In order to avoid such an approach with the search for random mutations, which might trap the intermediate state of p19, we tried to align and discuss our *in vitro* findings in the context of a cellular environment.
As already mentioned, the four members of the INK4 family share a similar protein fold, consisting either of four (p15, p16) or five (p18, p19) ARs. Various mutations are known, which inactivate single members of the INK4 family leading to diverse types of cancer and therefore attribute them to tumour suppressor proteins. Although they appear structurally redundant and equally potent as inhibitors, a number of non-overlapping features have been described [111-113].
Conspicuous is the short half-life of p19 in the cell, which was determined to ~ 20 min. In contrast, the half-life of p16 ranges between four to six hours in cell lines, although the thermodynamic stability is strongly reduced compared to p19. It was shown, that the periodic oscillation of p19 during the cell cycle is controlled by the ubiquitin/proteasome dependent mechanism, which appears to be restricted to p19 within the INK4 family. Lysine 62, located in the second AR of p19, was shown to be targeted by ubiquitination [114].
Analysis of further posttranlational modifications revealed a different phosphorylation pattern for the INK4 members. While no phosphorylation was seen for p15 and p16, p18 showed a detectable and p19 a strong phosphorylation signal. Single and double phosphorylated p19 species were isolated and phosphorylation sites were assigned to S66 and S76 [111]. Since the specific kinase that phosphorylates p19 *in vivo* is yet not known, we mimicked the phosphorylation sites by glutamate mutations to study the role of phosphorylation on stability,

kinetics and function. This artificial posttranslational mimic is widely used, because a negative charge at the right position seems often to be enough to approximate the function of the modified protein [115; 116].

Urea induced unfolding transitions of these mutants were monitored by tryptophan fluorescence (Trp86 was a suitable probe for monitoring p19 folding; see chapter 2.2). Analysis of these data revealed that all mutants with a glutamate at position 76 were strongly destabilized compared to the wild type protein, whereas the glutamate mutation on position 66 had less impact (Fig. 9A-D). Furthermore, S76E containing mutants now displayed three-state behaviour under equilibrium conditions, clearly showing, that the earlier detected hyperfluorescent intermediate state becomes significantly populated.

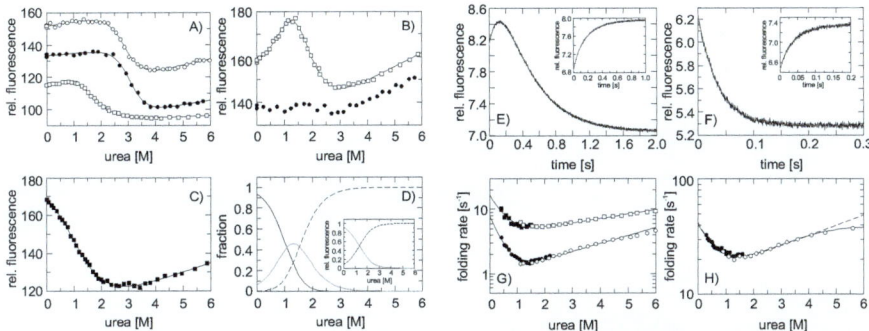

Fig. 9. (A-D) Urea-induced unfolding of p19^{INK4d} mutants monitored by tryptophan fluorescence. Transition curves of p19 F86W S66E (○), p19 F86W S76E (□) and p19 F86W S76A (●) at an emission wavelength of (A) 325 nm and (B) 375 nm at 15 °C. (C) Unfolding of p19 F86W S76E/S66E (■) at 37 °C. Solid lines in (A-C) represent the least square fit of a two-state or three-state model. (D) Calculated equilibrium populations for the p19 F86W S76E mutant of the native N (black line), intermediate I (grey line) and unfolded state U (dotted black line) according to the global analysis of the fluorescence equilibrium data at 15 °C. Inset shows the population profile for the same mutant at 37 °C. (E-H) Single mixing unfolding and refolding kinetics of p19 F86W S76E/S66E detected by stopped-flow fluorescence. Experimental data are plotted in black and fits in grey. Unfolding was initiated by a rapid change from 0 M to 3.2 M urea at 15 °C (E) and 37 °C (F) and can be best fitted by a double or single exponential function, respectively. Insets show fast refolding kinetics of the latter p19 mutant from 4.4 M to 0.4 M urea at the given temperature. The slowest refolding phase caused by prolyl *cis/trans* isomerisation is omitted for clarity. (G, H) Urea dependence of apparent folding rates of p19 F86W S76E/S66E monitored at 15 °C and 37 °C. Closed symbols (●,■) represent refolding experiments, open symbols (○,□) unfolding experiments.

Biophysical data pointed out, that the S76E mutation strongly destabilized the native state, whereas the stability of the intermediate state is marginally affected. The *m*-values for all mutants are similar within experimental error, suggesting that the phosphorylation mimic does not change the folding mechanism, but the stability (for details see subproject B). By measuring the stability at body temperature (37 °C) the picture changed. Equilibrium folding of S76E containing p19 mutants was now simplified to a two-state mechanism (Fig. 9C). The fluorescence was quenched upon addition of urea, indicating the transition from the hyperfluorescent intermediate to the unfolded state. This assumption is further confirmed by kinetic experiments. The characteristic biphasic "overshoot" kinetic with the hyperfluorescent intermediate state is still observable at 15 °C, while all kinetics at 37 °C are monophasic. Kinetics at 37 °C could be assigned to the I to U transition (Fig. 9F,H).

The high population of the intermediate state under equilibrium conditions, caused by the phosphorylation mimic, allowed a further characterization of this state by NMR spectroscopy. At 15 °C more than 82 percent of the backbone assignment of p19 could be directly transferred to the S76E mutant. ^{15}N-TROSY-HSQC spectra were recorded between 15 °C and 40 °C. Thereby native cross-peaks of AR 1 and 2 completely vanished at 37 °C, whereas AR 3-5 remained folded with native-like chemical shifts (Fig. 10). Thus, these high resolution NMR data confirmed the proposed scaffold function of AR 3-5 for the less stable AR 1 and 2.

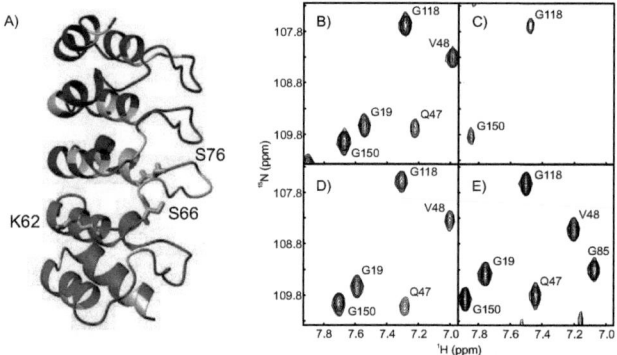

Fig. 10. (A) Schematic representation of the crystal structure of p19^{INK4d} (1bd8.pdb from the pdb) S66, S76, and K62 are indicated by a stick illustration of the side chains. Residues of the phosphorylation mimic mutant p19 F86W S76E with native chemical shift at 37 °C are color coded in blue, while residues in red lost the native structure at body temperature. Indicated in grey are proline residues and residues which could not been evaluated due to signal overlap or missing assignment. (B-E) Sections of ^{15}N-TROSY-HSQC spectra of p19 F86W S76E (B,C) and p19 wild type (D,E) at 15 °C (B,D) and 37 °C (C,E). Native cross-peaks of AR 1-2 of p19 F86W S76E vanished at 37 °C, whereas AR 3-5 still display native chemical shifts under these conditions. ^{15}N-TROSY-HSQC spectra of the wild type protein do not change between 15 °C and 37 °C.

The introduction of a negative charge, localized between AR 2 and 3 is sufficient to destabilize the native state in such manner, that just the intermediate state is populated at 37 °C. Since the modification of S76 by addition of a negative charge also occurs in the cell, the question about their role in a cellular context arises. A plausible explanation would be a local unfolding event of AR 1 and 2 after phosphorylation. This would allow the ubiquitin ligase to access lysine 62 more easily and finally target p19 to the proteasome. Because the specific ubiquitin ligase is not known, we tried to ubiquitinate p19 by using HeLa cell lysate extracts to test this hypothesis. Ubiquitination assays were negative for p19 wild type, while p19 S66E/S76E showed the strongest ubiquitination signal *in vitro*.

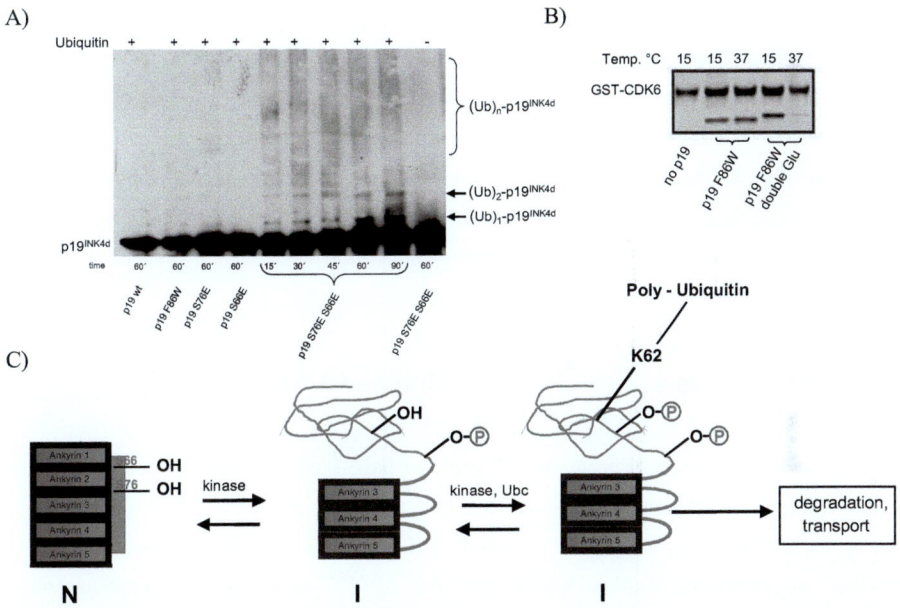

Fig. 11. (A) Ubiquitination of p19 *in vitro* by HeLa cell lysates requires at least two phosphorylation sites. The positions of unmodified and ubiquitinated p19 are indicated. Reaction mix was resolved by SDS-PAGE (4-20%), and visualized by autoradiography using specific p19 antibodies. (B) SDS-Page analysis of the pull-down assay of p19 F86W and p19 F86W S76E/S66E at 15 °C and 37 °C. Affinity of the double phosphorylation mimic mutant is strongly reduced at body temperature. (C) Functional folding model of p19: Phosphorylation of S76 leads to unfolding of the functional ARs 1-2, while ARs 3-5 remain folded. A second phosphorylation at S66 is necessary to allow efficient ubiquitination at K62 and subsequent degradation *via* the proteasom.

Although mutation S66E has a minor effect on the stability, it seems to be important for ubiquitination. Single mutants containing either a glutamate at position 66 or 76 were hardly targeted by ubiquitination. Thus, it is not surprising, that mainly double phosphorylated p19 molecules were found in cell lines. These data allowed us to build up a model, where phosphorylation of S76 unfolds the first two ARs of p19, but a second phosphorylation on S66 is necessary for efficient ubiquitination (Fig. 11).

These findings are in good agreement with literature data. It was shown earlier, that ubiquitination of cyclin E (in the same signal cascade) requires post translational phosphorylations close to the ubiquitination site to ensure efficient binding for the ubiquitin ligase [117; 118]. Our model has to be verified by future *in vivo* studies, because up to now all conclusions were drawn from *in vitro* experiments. But one question remains open: Does phosphorylation of p19 influence binding to CDK4 and 6 and thereby directly affecting the key step of the G1 to the S phase transition in the cell cycle?

We tested binding behaviour of the double glutamate mutant to CDK6 at 15 °C and 37 °C by GST pull down assays and analytical gel filtration. At low temperature, there was no obvious difference in CDK6 binding properties between wild type p19 and the glutamate mutant. At 37 °C, however, binding of p19 S76E/S66E was strongly reduced compared to wild type, but not completely abolished. Based on these experiments only, further conclusions would be highly speculative. Therefore, additional experiments will be necessary to describe the full role of p19 phosphorylation *in vivo*.

2.4 Ankyrin Repeat Proteins of Archaea - tANK

Folding studies on naturally occurring AR proteins have been until now focused only on eukaryotic proteins. This makes sense in a way that more than 90 percent of all AR proteins are found in eukaryotes. But to test the validity of a possible common mechanism of AR folding, we performed a Blast search with the $p19^{INK4d}$ sequence as a template on evolutionary much older archaeal organisms. An open reading frame of similar length and with less than 25 percent sequence identity to $p19^{INK4d}$ was identified in *Thermoplasma volcanium* [119].

Structure prediction programs classified this sequence as a putative AR protein. This organism was grown at 60 °C under anaerobic conditions for two weeks. Extracted DNA was used to amplify the desired gene by flanking primers and cloned in a pET-vector system. The protein was expressed as soluble form in high amounts and purified to homogeneity. Concentrated protein samples were subjected to various crystallization screens. After optimization of crystallization conditions, crystals diffracted up to 1.65 Å. The determined X-ray structure showed that this archaeal AR protein (tANK) indeed folds into five sequentially arranged ARs with an additional helix at the N-terminus. Single ARs of tANK show the typical characteristics found for ARs from eukaryotic proteins (Fig. 12).

Since a Blast search using the tANK sequence as starting sequence, mainly comprises archaeal homologues, a horizontal gene transfer seems to be unlikely. So, AR proteins must have been present throughout evolution. But what about folding? Did evolution select for proteins with high cooperativity as recently suggested by Watters et. al. [120]? Do we expect different folding behaviours of evolutionary older proteins compared to their eukaryotic homologues?

tANK turned out to be an ideal protein target for folding studies, due to its large fluorescence change between native and unfolded state and high reversibility. Trp 71 and Trp 104 located in AR 2 and 3 proved to be excellent probes to follow the transition curve. Upon unfolding, the fluorescence of the native state is strongly quenched and the maximum of the spectrum is shifted to a higher wavelength from 333 nm to 355 nm. At medium concentrations of GdmCl, however, an intermediate state gets populated with a quenched maximum, still at 333 nm (Fig. 12B).

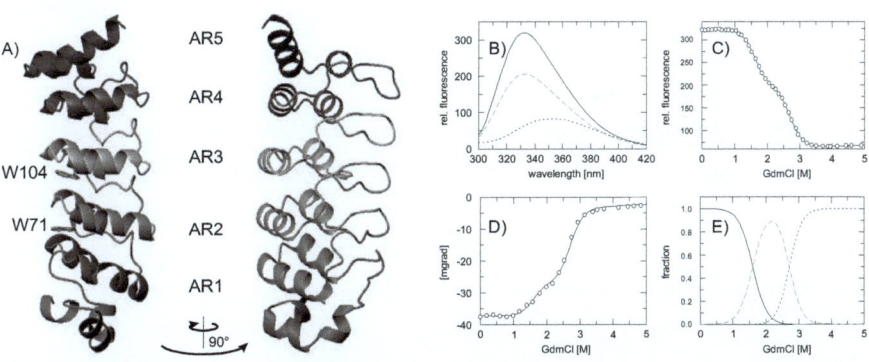

Fig. 12. (A) Schematic representation of the structure of the thermophilic ankyrin repeat protein tANK (2RFM.pdb in Protein Data Base). Five ARs (AR1 – AR5), each comprising a loop, a β-turn, and two sequential α-helices form the elongated structure, extended by an α-helical N-terminus. Side chains of the wild type fluorescence probes W71 and W104 are indicated as sticks. (B-E) GdmCl-induced unfolding of tANK monitored by fluorescence and CD spectroscopy. (B) Fluorescence spectra of tANK at 0 M (black line), 2 M (broken grey line), and 5 M (dotted black line) GdmCl after excitation at 280 nm. GdmCl induced unfolding transitions monitored by (C) fluorescence at 335 nm and (D) CD at 222.6 nm. Solid lines in (C), (D) represent the least square fit of a three-state model. (E) Calculated equilibrium populations of the native N (black line), intermediate I (broken grey line) and unfolded state U (dotted black line) according to the global analysis of fluorescence and CD equilibrium data.

A detailed analysis of the unfolding curves of tANK monitored by either fluorescence or CD spectroscopy (Fig. 12B-E), according to a three-state model revealed that the intermediate state gets populated up to 90 percent under equilibrium conditions ($\approx$ 2.1 M GdmCl). Cooperativity parameters for both transitions are very similar, and in the same range as seen for p19^{INK4d} (m_{NU} = 19.2 kJ·mol^{-1} M^{-1} for p19^{INK4d} measured in GdmCl). Analytical ultracentrifugation confirmed the monomeric state of the native and intermediate form of tANK.

Besides, folding kinetics were measured by stopped-flow fluorescence spectroscopy. Unfolding under fully denaturing conditions (> 3 M GdmCl) is fast and best described by a biphasic process, with 50 percent of the whole unfolding amplitude contributing to each reaction (Fig. 13A). The refolding reaction starting from fully unfolded molecules was best described by three exponential functions. But the fastest of these reactions accounts for more than 85 percent of the whole amplitude and is the only rate constant depending on the GdmCl concentration (Fig. 13B).

The two slow refolding reactions could be assigned to isomerization processes in the unfolded state. Nevertheless, a very fast refolding phase in the range of 100 s^{-1} and 1000 s^{-1} could not be observed in this experiment, although the entire refolding amplitude was detectable. These findings can be explained by the sequential folding mechanism U $\leftrightarrow$ I $\leftrightarrow$ N, which was already observed for p19^{INK4d} folding, with the formation of the intermediate state as rate-limiting step in the refolding reaction.

Kinetics derived from single mixing experiments can only directly monitor reactions before the rate-limiting step (exceptions were discussed in chapter 2.2). This explains the absence of a very fast refolding phase, when refolding is initiated from the unfolded state. To verify this hypothesis, unfolding and refolding reaction were initiated from the intermediate state. As proposed, refolding kinetics were very fast (inset Fig. 13B). The resulting rate constants filled the missing gap of the Chevron plot (Fig. 13C). Simultaneously, this fast folding reaction was assigned to the I $\rightarrow$ N transition. Unfolding kinetics of the intermediate state matched with the slow unfolding reaction observed in unfolding reactions starting from the native state. Therefore, we allocated these rates to the slow reaction between the intermediate and the unfolded state.

Summary and Discussion

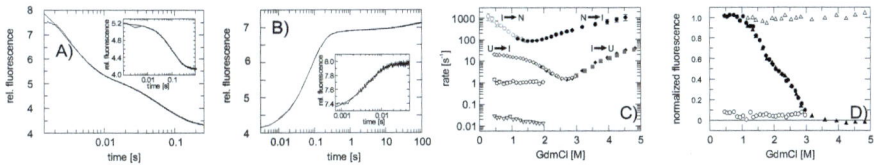

Fig. 13. (A, B) Single mixing unfolding and refolding kinetics of tANK detected by stopped-flow fluorescence. Experimental data are plotted in black and fits in grey. (A) Unfolding was initiated by a rapid change from 0 M to 4 M GdmCl and can be best fitted by a double exponential function. (B) Refolding was initiated by rapid dilution from 5 M to 0.9 M GdmCl and follows a sum of three exponentials. Refolding (inset A) and unfolding traces (inset B) starting from GdmCl concentrations where the intermediate is highly populated (1.7 M GdmCl for refolding and 2.6 M GdmCl for unfolding) can be best described by a single exponential function. (C) GdmCl dependence of apparent folding rates of tANK monitored at 15 °C, pH 7.4. Closed symbols (●,◆) indicate unfolding experiments, open symbols (◇,○,▽) indicate refolding experiments. Grey symbols (○,■) represent folding rates which result from unfolding and refolding kinetics starting from the intermediate state. (D) Start and end point analysis of the kinetic experiments. End points of unfolding (▲) and refolding (●) reactions follow fluorescence equilibrium data. Start point (○,△) analysis reveal no obvious burst-phase.

Analysis of the folding and unfolding kinetics of the AR proteins p19^{INK4d} and the evolutionary older archaeal homolog tANK revealed the same folding mechanism for both proteins, with the formation of the intermediate state as rate-limiting step.

To receive information on the structural properties of intermediate tANK, we monitored the GdmCl transition by NMR (Fig. 14). For that purpose, more than 80 percent of the backbone amide protons of the native state were assigned using standard 3D experiments. Native cross-peaks were followed by a series of 19 2D ^{15}N-TROSY-HSQC spectra at various GdmCl concentrations. 66 out of 185 possible native amide cross-peaks could be followed during the entire transition without any overlap of cross-peaks from I and U. They can be divided into two classes, following the two transitions seen from fluorescence and CD data. Cross-peaks of N-terminal AR 1 and 2 show a transition midpoint close to 1.6 M GdmCl and completely vanished at 2.1 M GdmCl, where the intermediate state is maximally populated. In contrast, residues of AR 3-5 still showed native chemical shifts at 2.1 M GdmCl and unfold cooperatively with a transition midpoint of ≈ 2.6 M GdmCl (Fig. 14B).

Detailed analysis allowed the assignment of the two transitions observed by optical methods to the respective residues in tANK. Amide protons of AR 1 and 2 followed the decay of the native population in accordance with fluorescence and CD data. However, transition curves of residues from AR 3-5 could be best described by the sum of the native and intermediate population. This demonstrates that AR 3-5 show native-like chemical shifts in the intermediate state. Furthermore, twelve additional peaks could be directly assigned to the intermediate state indicating that not all N-terminal residues sense a completely unfolded environment (Fig. 14C). The courses of intermediate cross-peaks corresponded well with the intermediate population calculated from the fluorescence data.

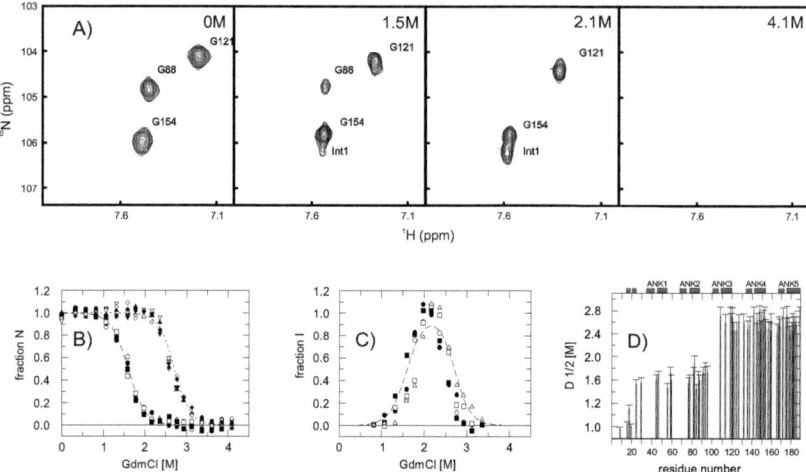

Fig. 14. GdmCl induced unfolding transitions of tANK monitored by NMR. ^{15}N TROSY-HSQC spectra were recorded between 0 M and 4.2 M GdmCl. (A) Sections of ^{15}N TROSY-HSQC spectra of tANK show the disappearance of native cross-peaks at low (e.g. G88) and high (e.g. G121) denaturation concentrations. Transiently appearing cross-peaks of the intermediate state are labeled with Int. (B) Normalized cross-peak volumes of backbone amides assigned to the native state at 0 M GdmCl. E45, D60, L78, G88, V91 of AR 1-2 follow the decay of the native state population derived from the fluorescence and CD data (broken line). G109, E119, G142, L153, A189 of AR 3-5 follow the sum of the native state and intermediate state population derived from the fluorescence and CD data (dotted lines). (C) Additional, transient cross-peaks which do not heavily overlap with peaks from the native or denatured state agree with the intermediate population (broken line) resulting from fluorescence and CD data. (D) Midpoints of denaturation profiles of 66 out of 185 possible amide cross-peaks show that the two N-terminal AR are by 1 M GdmCl less stable compared to C-terminal three repeats.

The folding analysis of the human CDK inhibitor p19^{INK4d} and the archaeal tANK protein revealed that both proteins fold *via* an on-pathway intermediate to the native state. The formation of the intermediate state is rate-limiting in the folding reaction, therefore providing a scaffold function for the less stable repeats. Although the cellular function of tANK is not known, it is allowed to speculate, that the N-terminal repeats host the functional binding interface. Furthermore, it has been shown, that partially unfolded ARs of single proteins fold into a characteristic AR upon target binding [121]. This might be also the case for the N-terminal part of tANK, which was designated as N-terminal helix in this work.

Although these two analyzed proteins are far separated in evolution, their native structures resemble each other and they display the same highly coopertive folding mechanism, including a similar folding intermediate based on the number of folded ARs. These findings lead to the following conclusions: AR proteins have already evolved quite early in evolution their characteristic properties and functions, but the organisms at that time did not take advantage of this system, because they were far less complex than eukaryotic systems. Hence, with increasing cellular complexity, the organisms had to adopt quite efficiently and rapidly their protein repertoire to the new environmental conditions. This was possible with the modular architecture of repeat proteins, because simple mutations, deletions, insertions, or duplications of these genes result in a large variety of folded and functional proteins, which can adapt easily their protein interfaces for specific functions. Higher eukaryotes took advantage of this system and the high abundance confirms their success.

2.5 The Yin and Yang of Repeat Protein Folding

The extended architectural simplicity of linear repeat proteins is a major advantage in studying protein folding. Long range topologies of globular proteins prevent dissection of energetics to different structural elements, which is required to construct energy landscapes. Repeat proteins are the solution to this problem, because sequence distant contacts are absent as visualized in contact maps (Fig. 15) [122]. Hence, distant regions of repeat proteins do not directly impact each other's structure. Therefore, it should be possible to study the folding of different regions of repeat proteins independently, either by deleting repeats or structurally characterize partially folded states in which a subset of repeats remains structured.

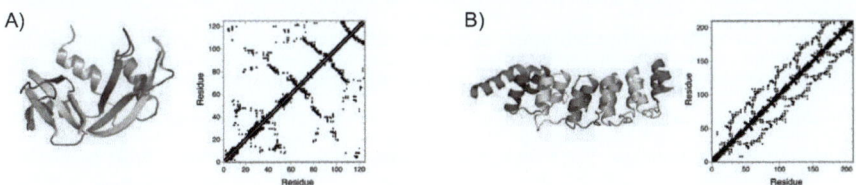

Fig. 15. Comparison of globular and repeat protein structures. (A) A globular protein used in protein folding studies (RNase A, 7RSA.pdb) is compared to (B) a naturally occuring repeat protein (Notch ankyrin domain, 1OT8.pdb, chain a). Structures are colour coded according to different secondary structural elements and repeats. Contact maps emphasize the lack of long range interactions in repeat proteins and regular patterns of structure elements in repeat proteins (after Kloss et al. [122]).

Various folding studies on natural and synthetic repeat proteins have been carried out and different conclusions were drawn. The Notch ankyrin domain was used to experimentally determine the energy landscape, by generating overlapping constructs containing subsets of the seven ARs of the *Drosophila* Notch receptor [123; 124]. The stabilities of each construct could be described as a sum of energy terms associated with each repeat and defined an equilibrium energy landscape. Nevertheless, deletion constructs of natural AR proteins should be handled with care. Nature has optimized N- and C-terminal repeats (so-called capping repeats) to prevent the exposure of hydophobic surfaces and increase solubility. Deleting these repeats might dramatically effect solubility, reversibility, and stability of the constructs, which makes the interpretation of energy terms in the context of the full length protein difficult. Capping repeats of natural AR proteins were therefore used for designed AR proteins, because synthetics AR proteins, consisting of tandem repeats derived from the consensus sequence are mainly unsoluble at neutral pH [63].

To overcome this problem tetratricopeptide repeats (TPRs) of variable length were designed based on a consensus sequence. In this case, the designed proteins contained different numbers of an identical repeated unit, without any optimized capping repeats. Kajander and coworkers have shown in a very elegant way, that folding of TPR proteins can be described by a classical one-dimensional `Ising model` [125]. Using this model, they were able to predict the stability of TPR proteins of different length. Furthermore, this model predicts the existence of different partially folded intermediate states under equilibrium conditions, but experimental evidence is missing up to known. Whether the use of an one-dimensional `Ising model` represents a new folding paradigm for repeat proteins remains open, because natural repeat proteins consist of a number of repeats different in sequence, with different stabilities and interaction properties towards adjacent repeats.

Our work on natural AR proteins clearly showed the existence of specific and not different intermediate states, which could be characterized at high resolution. Natural AR sequences can be very different among each other and hence display very different intrinsic stabilities. While AR 3-5 of p19 (and also tANK; data not shown) are able to fold independently, AR 1-2 in isolation remain unfolded although it has been shown that two ARs form the minimum folding unit of AR proteins. We should keep in mind, that these proteins fulfil a specific function in the cellular context and protein folding can be a (useful) direct or indirect tool to control functionality, localization, or life-time of the protein. The idea derived from our findings is that more stable AR repeats form a scaffold for the less stable but functional repeats. This might be a general mechanism for natural repeat proteins, each with a specific function according to different environments.

2.6 SlyD – Prolyl Isomerase and Folding Helper

In the first part of this thesis, the work focussed on structural properties of folding intermediate states. Therefore mutations were introduced to trap proteins in a partially folded state or NMR spectra in the presence of denaturant were recorded and analyzed, in order to fill the gap of structural information on folding intermediate states.

In the second part, we concentrate on enzymes, which prevent the accumulation of partially folded states by speeding up their folding reactions to the native state.

As mentioned in chapter 1.2, accumulation of folding intermediates raises the risk of misfolding and aggregation. To suppress this process and enhance productive folding, nature has evolved folding helper proteins. SlyD is one of them. It is an efficient peptidyl-prolyl *cis/trans* isomerase (PPIase) and shows in addition chaperone-like activities [126]. Already 15 years ago, SlyD was discovered in *E. coli* as a host factor for the ΦX174 lysis cycle [127-130].

Based on sequence comparisons, it was proposed that SlyD belongs to the FKBP family (one group of PPIases) [131], for which binding of the immunosuppressant FK506 is characteristic [132]. Indeed, the isolated protein was highly active in prolyl isomerase assays. Further research results revealed chaperone activity for SlyD in an ATP independent manner.

SlyD is probably one of the most obnoxious proteins in molecular biology because it contains a cysteine- and histidine-rich tail, which is suggested to act as metal binding domain. Hence, recombinant proteins, overexpressed in *E. coli* and purified *via* IMAC chromatography are almost always contaminated with SlyD. Although various structure groups claimed that they have solved the structure of SlyD by accident, so far no structure files have been deposited in the pdb databank. The protein attracted attention in the last few years, because it could be shown, that covalent fusion of aggregation prone proteins with SlyD modules strongly enhanced cytosolic expression and solubility. Therefore, it became a valuble tool in diagnostic biotechnology [56; 133].

Although this protein is well known, the mechanistic details of this enzyme are poorly understood. In addition, SlyD does not just consist of an FKBP domain, it also contains an 45 amino acid insertion in the flap region (according to the nomenclature of human FKBP12), close to the prolyl isomerase active site. The structure of SlyD from the thermophilic

Summary and Discussion

organism *Thermus thermophilus* (tSlyD), which shows more than 50 percent sequence identity to *E. coli* SlyD* (1-165), but lacks the C-terminal cysteine- and histidine- rich tail, was solved by X-ray crystallography (Fig. 16). The deletion of this C-terminal part was shown to have no influence on the prolyl isomerase and chaperone activity in *E. coli* SlyD*. tSlyD was also highly active in prolyl isomerase and chaperone assays. The crystal structure of tSlyD comprises a FKBP-like domain and the insertion in the flap folds into an autonomous domain (IF domain). Diffracting crystals of tSlyD were obtained under two different conditions in different space groups. Resulting structures of this two-domain protein showed variability regarding their domain orientation, while the overall shape of the domains was basically unchanged. Actually, the difference in domain orientation of these two structures is caused by a twist and short movement of the linker region, which orients the domains in closer proximity in one structure compared to the other (Fig. 16B-D). A detailed structure characterization is found in subproject D.

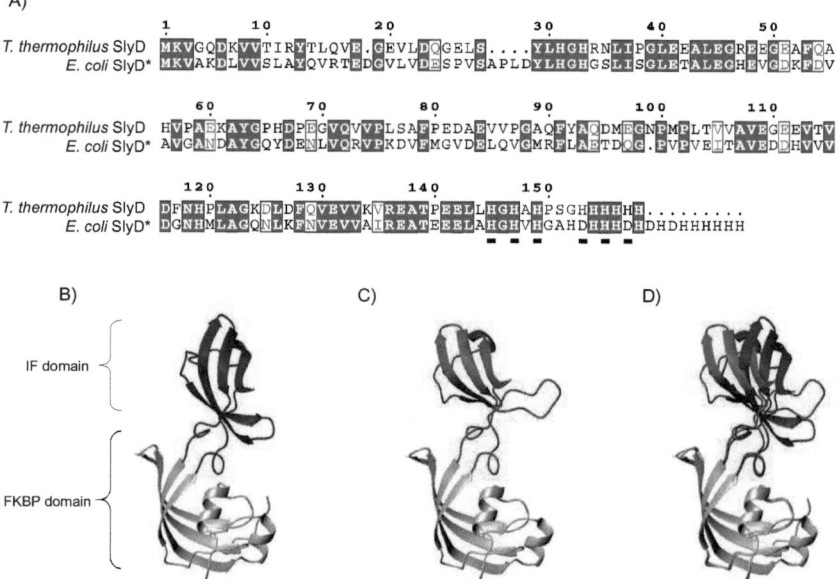

Fig. 16. (A) Sequence alignment of tSlyD and *E. coli* SlyD* (1-165). Identical and similar residues between these two proteins are boxed or coloured in red, respectively. Both constructs carry a C-terminal histidine tag. Histidine residues involved in Ni^{2+}-ion binding are underlined. (B,C) Schematic representation of the crystal structure of tSlyD (3CGM.pdb, 3CGN.pdb in Protein Data Base) derived from two different crystal forms (crystal form A (B), crystal form B (C). The FKBP and the IF domain are indicated. (D) Superposition of the crystal structures of tSlyD demonstrates the different orientation of the two domains towards each other. The C_α atoms of the FKBP domain were used for superimposition.

Summary and Discussion

Deletion mutants and binding studies of tSlyD with various unfolded substrates followed by NMR, ITC, and fluorescence clearly showed, that the chaperone activity is located in the IF domain, whereas the FKBP domain exhibits the expected prolyl isomerase activity. Binding is mainly mediated *via* hydrophobic interactions (Fig. 17).

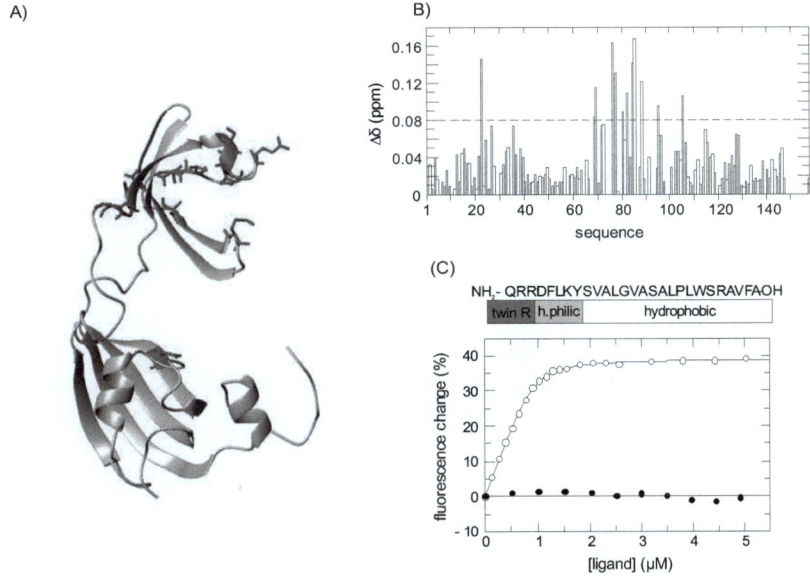

Fig. 17. Mapping the tSlyD chaperone binding interface by NMR titration and fluorescence spectroscopy. (A) Residues of tSlyD showing chemical shift changes $\Delta\delta > 0.08$ ppm upon binding to RCM-α-lactalbumin are colored in red. (B) Backbone chemical shift changes ($\Delta\delta$) of tSlyD upon binding to RCM-α-lactalbumin. (C) Sequence and schematic representation of the Tat-signalpeptide. Fluorescence increase of Trp 21 of the Tat-signalpeptide upon binding to tSlyD (○). tSlyDΔIF lacking the IF domain (●) did not bind.

Moreover, we could express both domains isolated in a folded and functional state and could directly assign the different functions (data not shown). The ^{15}N-HSQC spectrum of tSlyD is a linear combination of the spectra of the isolated domains. By using the NMR H/D exchange technique, we were able to probe the local stability of tSlyD under native conditions. Interestingly, amide protons of the FKBP domain, were, on average, 1000-fold more protected, than those of the IF domain. This proves, that the latter undergoes frequent local opening without unfolding the FKBP domain and is significantly less stable than the FKBP domain (for details see subproject D).

In binding studies, single expressed domains did not interact with each other, indicating that both domains are highly flexible and can adopt different orientations in the full length protein. This agrees well with our findings derived from the crystal structures of tSlyD and NMR structures of *E. coli* SlyD* (Weininger U., unpublished results). By using SAXS analysis, however, we were able to assign the most preferred domain orientation in solution to the structure of crystal form A, which shows a slightly kinked arrangement of the domains. Interestingly, scattering curves for tSlyD in complex with the Tat-signalpeptide were not significantly altered when compared to the free form (Fig. 18). These data do not support a model where substrate binding orients both domains constantly in direct proximity.

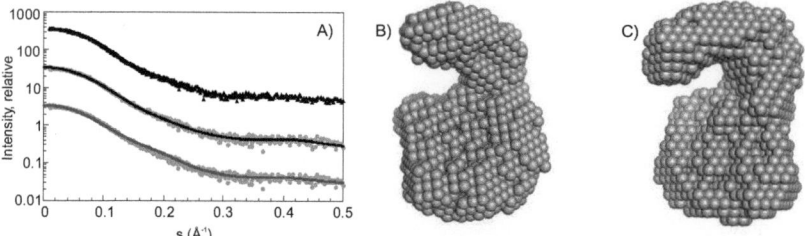

Fig. 18. SAXS analysis of tSlyD. (A) Experimental intensities for free tSlyD (light grey) and a tSlyD/Tat-signalpeptide complex (black triangles). The scattering profiles are displaced along the ordinate for better visualization. The fit to the scattering pattern computed from the crystal structures of tSlyD is shown in black (chi = 1.61) (crystal form A, middle) and grey (chi= 2.43) (crystal form B, bottom). *Ab initio* low resolution structure models of tSlyD (B) and the tSlyD-Tat-signalpeptide complex (C) calculated from the SAXS pattern. The balls represent the dummy atoms in the simulated annealing procedure to restore the models.

Nevertheless, the presence of the chaperone domain has an enormous impact on the activity of the prolyl isomerase. The catalytic efficiency towards proline limited refolding of ribonuclease T1 is 100-fold reduced, when the chaperone domain is absent in tSlyD (for details see subproject D). Therefore, a certain type of coupling of both domains must exist, to explain this dramatic activity change. This makes SlyD an interesting target to study because the combination of a catalytic domain with a chaperone domain is a common principle in nature. But in contrast to the most prominent member "trigger factor" [54], which also exhibits prolyl isomerase and chaperone activity, the architecture of SlyD is much simpler .

The influence of the IF domain on a catalytic domain was further demonstrated by the chimeric protein "Thermus BP12". This artificial protein consists of human FKBP12, but the original flap region (nine amino acid long), was replaced by the IF domain of tSlyD. The presence of the chaperone domain turned FKBP12 into an excellent catalyst of proline limited protein folding, with a 180-fold increased activity compared to FKBP12 only. This chimeric protein is even more active than various SlyD species, tested from different organisms.

But what are the mechanistic details? Are there any limiting factors? The simplest explanation for the increased activity of chaperone domains containing prolyl isomerases, of course, comes along with the additional binding site for a substrate close to the active site. This leads to an increased local concentration and thus higher activity.

The question remains: Is it that simple? The different orientations of both domains found in crystal and NMR structures suggest a swinging arm like mechanism, where substrates bound to the IF domain are translocated close to the active site of SlyD by domain rearrangement. Recent NMR dynamic studies on enzymes revealed, that the intrinsic dynamics of certain enzymes can be strongly coupled to the turn-over rate of their reactions [134-136]. To validate such a model for SlyD it will be of great interest but also very challenging to unravel, whether the domain dynamics of free SlyD, is somehow correlated to the turn-over rates of substrates. Indeed, we do not expect a direct correlation of domain dynamics with the turn-over rate, rather propose, that the domain dynamics display an upper limit of reachable turn-over rates. NMR R_2-dispersion experiments in combination with single molecule FRET spectroscopy will be the most promising tools for future experiments.

2.7 Bringing Your Curves to the BAR

Lipid membranes are fundamental for cellular structure. They can adopt many different shapes and even line out smaller vesicles [81; 82]. Clathrin mediated endocytosis is one of the key mechanism by which cells take up liquids or particles from the environment [83; 85; 137]. Changing the membrane shape often requires the formation of high curvature microdomains. Nature has evolved protein modules, which are able to induce and stabilize such membrane curvatures.

Within those proteins, BAR (Bin/Amphiphysin/Rvs-homology) domains have been identified throughout eukarya as important regulators of membrane remodelling processes [86; 87; 138; 139]. The BAR domains self associate into crescent shaped homodimers with a positively charged concave surface. The latter is suggested to bind to negatively charged membranes to drive and/or sense curvature. Incubation of BAR domains with liposomes reshapes them into tubules (a process known as "tubulation"), with a diameter, similar to the curvature of the BAR domain [87].

A variant of the BAR domain, the N-BAR domain contains an N-terminal extension with amphipathic character, which is predicted to fold into a helix upon binding to the membrane (helix-0) [87; 88; 94]. The ability to tubulate liposomes is strongly enhanced for N-BAR domains compared to BAR domains lacking helix-0. The insertion of amphipathic helices into one leaflet of the bilayer is proposed to be a general mechanism for the generation of membrane curvature. Experimental evidence for structure induction and insertion of this amphipathic helix was so far just derived from circular dichroism and electron paramagnetic resonance spectroscopy (EPR) [87; 88]. Although crystal structures of N-BAR domains are known, high resolution information on this N-terminal appendage is missing, because it is proposed to be disordered in solution and thereby unresolved in all structures.

We have chosen an NMR approach combined with molecular dynamics (MD) simulations to study function, dynamics, and structure of helix-0 of the human Bin1/Amphiphysin II BAR domain in detergent and lipid environments. The N-terminal BAR peptide (residues 1-33) was recombinantly expressed and initially analysed by CD spectroscopy. The isolated N-BAR peptide was unstructured in aqueous solution but adopted a helical structure when bound to liposomes or micelles (Fig. 19). The CD spectra of the N-BAR-peptide in liposomes, DPC, or SDS micelles were mainly identical, indicating a similar structure under these conditions.

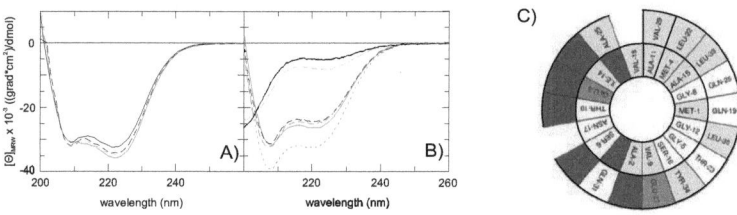

Fig. 19. Far-UV CD spectra of the BAR domain and the N-terminal N-BAR peptide in various solvent environments. (A) CD spectra of the N-BAR (residues 1-241) (solid line) and the BAR (residues 32-241) domain (dashed line) of human amphiphysin II in the presence and absence of brain lipid liposomes (grey and black respectively). Structure induction upon binding to liposomes is only seen for the N-BAR domain (solid grey line), indicated by a significant signal decrease at 222 nm. (B) The N-BAR peptide is unstructured in solution (solid black line). In the presence of liposomes (dashed dark grey line), SDS (solid grey line) or DPC (black dashed line) micelles, the peptide becomes structured. In the presence of OG (dashed light grey line) micelles however, no structure induction is observed. A CD spectrum recorded in 60 % TFE (dotted line) shows the highest helical content. (C) Helical wheel diagram for the N-BAR peptide. The amino acid sequence (three letter code) is plotted clockwise. Hydrophobic residues are shown in grey boxes and positively and negatively charged residues in blue and red, respectively.

But the type of detergent had strong influence on the interaction properties with the peptide. Non-ionic OG micelles as membrane mimic did not induce any secondary structure in the peptide. This implied, that binding and structure induction is not just driven by hydrophobic but also electrostatic interactions. This fact is further supported by MD simulations which were carried out on the N-BAR peptide in SDS and DPC micelles. In each case, the peptide migrated to the surface of the micelle, as expected for an amphipathic helix (Fig. 21A-D). Analysis of simulations showed, that cationic residues of the peptide strongly interacted with anionic head groups, and that the N-BAR peptide embedded deeper in the DPC micelle compared to SDS (for details see subproject E).

A comparison of ^{15}N-HSQC spectra of the N-BAR peptide in solution, SDS, and DPC micelles validated a defined secondary structure induction in lipid-like environment compared to aqueous solution (Fig. 20A,B). Structure calculation revealed an α-helical conformation for residues 8-34 in SDS and 10-34 in DPC micelles, which is consistent with CD data (Fig. 20C,D). The structured part of the N-BAR peptide clearly showed amphipathic character with negatively charged side chains on the convex side and hydrophobic side chains on the concave side.

The disordering of the N-terminal part was confirmed by ^{15}N-heteronulear NOE (hNOE) measurements, which monitored picoseconds to nanoseconds dynamics. The folded part of the N-BAR peptide is quite rigid in this time regime (hNOE > 0.5), typical for folded structural elements, whereas N-terminal residues displayed hNOE values close to zero or negative. The hNOE of an extended N-BAR peptide in DPC micelles with 44 amino acid dropped after K35 towards the C-terminus, indicating that the amphipathic helix ends at position 35 and following residues form the linker to helix-1 of the BAR-domain (Fig 20E-H).

Summary and Discussion

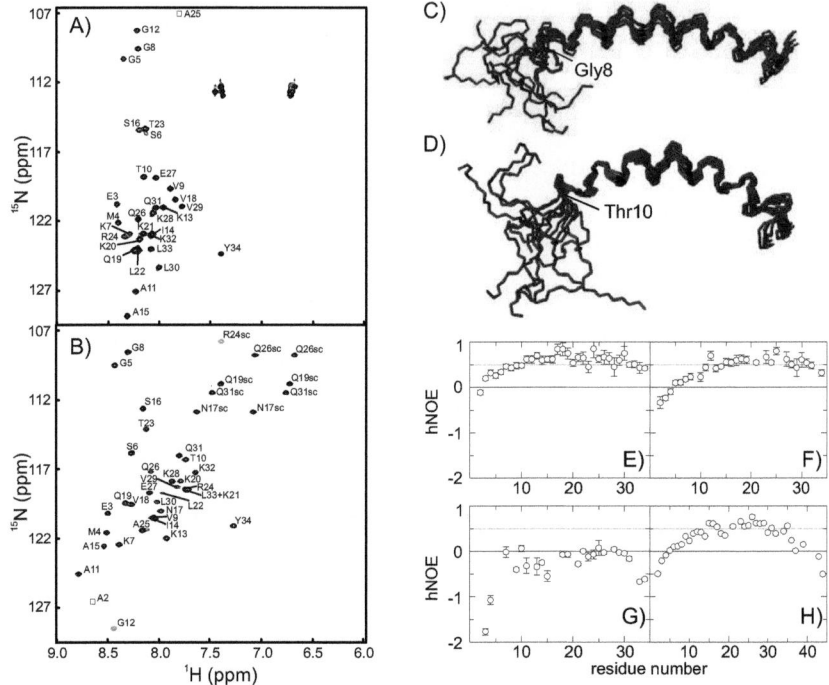

Fig. 20. 2D ^{1}H-^{15}N HSQC spectra of the N-BAR peptide (A) in aqueous solution, and (B) bound to DPC micelles. The assigned cross-peaks of the backbone amides are labeled using the one-letter amino acid code and the sequence position. Boxes indicate resonance signals, which show cross-peak intensities below the plotted contour level. (C,D) Structure ensembles of the N-BAR peptide backbone bound to detergent micelles at 25 °C: Ten lowest energy structures in (C) SDS micelles and (D) DPC micelles. Starting point of the amphipathic helix is indicated. (E-H) ^{1}H-^{15}N heteronuclear NOEs of the N-BAR peptide in (E) SDS, (F) DPC and (G) aqueous solution. (H) hNOE values of the extended N-BAR peptide (1-44 residues) in DPC micelles.

A fast ms-amide proton exchange experiment exposed residues with an increased solvent accessibility. N-terminal polar and charged residues showed a pronounced signal change during the experiment, while amide protons of hydrophobic residues did not exchange at all (Fig. 21E-G.). Together with the findings from MD simulations we drew the following picture: The N-BAR peptide embeds on the surface of the micelle, with a disordered N-terminal region exposed to the solvent, whereas hydrophobic residues are buried in the micelle.

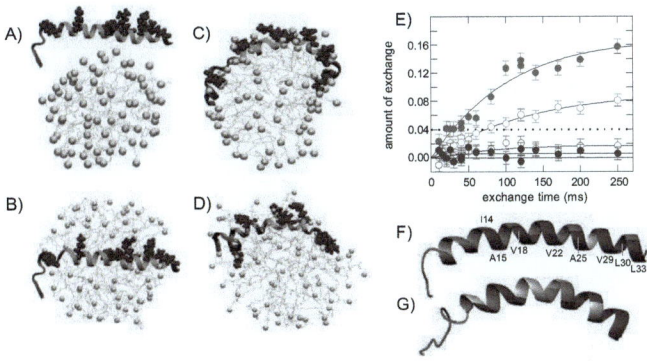

Fig. 21. (A-D) Snapshots of MD simulations at the beginning (left) and end (right) of 60 ns simulations of the peptide/micelle systems performed in SDS (A,C) or DPC (B,D) micelles. α-helical regions of the peptides are presented in green, non-helical in grey, and positively charged side chains in blue. The negatively charged sulfur of SDS and phosphorous of DPC are yellow, and the acyl chains are light blue. The N-BAR peptide is initially positioned inside or outside a micelle. Water and ions are omitted for clarity. (E-G) NMR experiment to detect fast exchanging amide protons (MEXICO) of the N-BAR peptide bound to SDS and DPC micelles. Fast amide proton exchange was followed on a residue by residue level. (E) Exchange curves in SDS micelles are shown for T23 (closed red symbols), S16 (open red symbols), L33 (closed blue symbols) and V29 (open blue symbols). Fast exchanging amides are colored in red. Amide protons, which did not exchange within the timescale of the experiment (below dashed line) are colored in blue. Exchange curves for residues in grey could not been evaluated due to signal overlap or low signal intensity. This color code was assigned to ribbon representation of the lowest energy NMR structure of the N-BAR peptide in (F) SDS and (G) DPC micelles.

We tested the functionality of the N-BAR peptide in the context of the full length BAR domain and in isolation by membrane tubulation and fusion assays (Fig. 22). The human amphiphysin N-BAR domain was able to constrict liposomes into tubules as shown by electron microscopy. A deletion mutant, lacking the N-terminal appendage (BAR), had no significant influence on the liposome morphology. Along with extensive vesiculation, the isolated N-BAR peptide showed also tube formation, but with diverse diameters compared to the full length protein. Membrane fusion experiments using fluorescence FRET spectroscopy demonstrated that helix-0 is highly fusogenic in isolation or in context of the full length protein (for details see subproject E). The BAR domain itself was not able to fuse liposomes (Fig. 20).

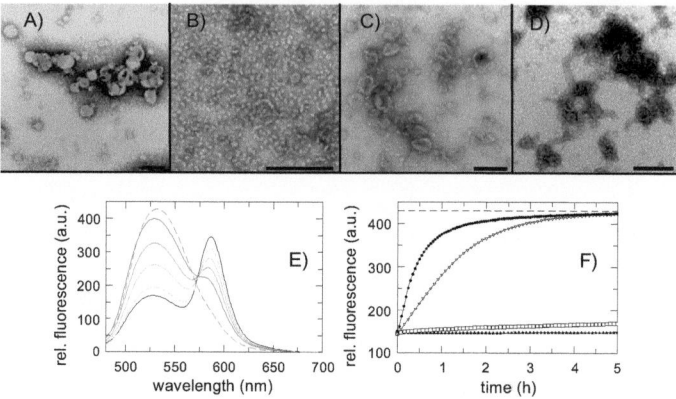

Fig. 22. Electron micrographs of liposome tubulation by (B) human amphiphysin N-BAR, (C) BAR and (D) the N-BAR peptide (length scale, black bar, 200 nm). Untreated liposomes are shown in (A). Fluorescence emission spectra (E) from mixed liposomes in the absence (black curve) and presence of the N-BAR domain at various time points (8 min, 28 min, 62 min, 225 min, light to dark grey) and 1 % Triton X-100 (for total donor fluorescence, dashed line). (F) Time dependent increase of donor fluorescence at 530 nm upon membrane fusion in the presence of N-BAR (●), BAR (□) and the N-BAR peptide (▽). Fluorescence change caused by spontaneous liposome fusion is negligible (▲) and at maximum in 1 % Triton X100 (dashed line).

The combination of experimental and theoretical techniques shed light on the structural characteristics of helix-0 of the human Bin1/Amphiphysin II BAR domain. High resolution structures confirmed the predicted amphipathic character of helix-0, but revealed an unstructured and solvent exposed N-terminal region. Activity assays highlighted the importance of helix-0 for the BAR domain, because it strongly increases the affinity of the N-BAR domain to the lipid bilayer.

Various mechanisms of membrane curvature are currently discussed. Based on our results we strongly favour the scaffold mechanism, which assumes that the intrinsic curvature of the BAR domain forces the membrane into bended shape. Helix-0 increases the affinity of the BAR domain to the membrane, but is not able to curve membranes into a uniform shape. Although helix-0 is able to drive membrane curvature, we do not expect a defined curving generation of the membrane prior to the interaction with the BAR domain.

3. Abbreviations

AR, ANK	Ankyrin Repeat
ATP	Adenosine-Tri-Phosphate
BAR domain	BAR (Bin;Amphiphysin;Rvs) domain
N-BAR domain	BAR domain with N-terminal amphipathic helix
F-BAR domain	Fer-CIP4 homology BAR domain
BAR peptide	34 amino acid peptide of residues 1-33 of the human Bin1/amphiphysin II BAR domain with a C-terminal tyrosin residue
CDK4, CDK6	Cyclin Depdendent Kinases 4 and 6
CD	Circular Dichroism
D	Denaturant
$[D]_{½}$	midpoint of unfolding transition
DPC	Dodecyl-Phospho-Choline
E. coli	*Escherichia coli*
EPR spectroscopy	Electron Paramagnetic Resonance spectroscopy
FKBP	FK506 Binding Protein
FRET	Förster Resonance Energy Transfer
GdmCl	Guanidinum chloride
GST	Glutathione *S*-Transferase
G1-phase	G_1 phase is a period in the cell cycle during interphase, after cytokinesis and before the S phase
ΔG^0_{ij}	free enthalpy of unfolding in the absence of denaturant between states i and j
helix-0	amphipathic helix of the human Bin1/amphiphysin II BAR domain
H/D exchange	Hydrogen-Deuterium exchange
His-tag	polyhistidine-tag for affinity purification
hNOE	heteronuclear Nuclear Overhauser Effect
HSQC	Heteronuclear Single Quantum Coherence
INK4	Inhibition of CDK4
IF domain	"inserted flap" domain
I	Intermediate state

Abbreviations

I-assay	kinetic assay to determine the population of intermediate molecules under equilibrium conditions
IMAC	Immobilized Metal Affinity Chromatography
ITC	Isothermal Titration Calorimetry
k_{app}	apparent rate constant
k_{ij}	microscopic rate constant for the reaction from i to j
m	cooperativity parameter for the unfolding reaction
m_{ij}	kinetic cooperativity parameter for the reaction i to j
MD simulation	Molecular Dynamic simulation
N	Native state
N-assay	test for native molecules in a kinetic double jump experiment starting from the unfolded state
NMR	Nuclear Magnetic Resonance
NOE	Nuclear Overhauser Effect
OG	Octyl-β-D-Glucopyranoside
P	Protection factor
PAGE	Poly-Acrylamide-Gel Electrophoresis
pdb	protein data bank
p19	cyclin dependent kinase inhibitor p19^{INK4d}; molecular weight of 19 kDa
p19 F86W	p19 protein with a tryptophan residue at position 86 instead of a phenylalanine
p19 AR1-4	deletion construct of p19 consisiting of AR 1 to 4
p19 AR3-5	deletion construct of p19 consisiting of AR 3 to 5
p15, p16, p18	Cyclin dependent kinase inhibitors p15, p16, p18
RP-HPLC	Reversed Phase – High Performance Liquid Chromatography
S-phase	synthesis phase, is a period in the cell cycle during interphase, between G1 phase and the G2 phase
SAXS	Small-Angle-X-ray Scattering
SDS	Sodium-Dodecyl-Sulfate
SH2 domain	Src homology 2 domain
SH3 domain	Src homology 3 domain
SlyD	product of the *slyD* (sensitive-to-lysis) gene

Abbreviations

E. coli SlyD*	1-165 fragment of SlyD
E. coli SlyD* ΔIF	variant of SlyD* in which the IF domain is replaced by the flap of human FKBP12
tANK	thermophilic ankyrin repeat protein
Tat pathway	twin arginine translocation pathway
TFE	trifluorethanol
Thermus BP12	variant of FKBP12 in which the flap was replaced by the IF domain of tSlyD
TPR repeats	tetratricopeptide repeats
tSlyD	SlyD protein of the organism *Thermus Thermophilus*
tSlyD ΔIF	variant of tSlyD in which the IF domain is replaced by the flap of human FKBP12
TROSY	Transverse Relaxation-Optimized Spectroscopy
U	Unfolded state
U-assay	test for unfolded molecules in a kinetic double jump experiment starting from the native state
wt	wild type

4. References

1. Anfinsen, C. B. (1973). Principles that govern the folding of proteins chains. *Science* **181**, 223-230.
2. Yon, J. M. (2002). Protein folding in the post-genomic era. *J Cell Mol Med* **6**, 307-27.
3. Fersht, A. R. & Daggett, V. (2002). Protein folding and unfolding at atomic resolution. *Cell* **108**, 573-82.
4. Gianni, S., Ivarsson, Y., Jemth, P., Brunori, M. & Travaglini-Allocatelli, C. (2007). Identification and characterization of protein folding intermediates. *Biophys Chem* **128**, 105-13.
5. Dobson, C. M. (2003). Protein folding and misfolding. *Nature* **426**, 884-890.
6. Brockwell, D. J. & Radford, S. E. (2007). Intermediates: ubiquitous species on folding energy landscapes? *Curr Opin Struct Biol* **17**, 30-7.
7. Levinthal, C. J. (1968). Are there pathways for protein folding? *J. Chim. Phys.* **65**, 44-45.
8. Dobson, C. M., Sali, A. & Karplus, M. (1998). Protein folding: A perspective from theory and experiment. *Angew. Chem. Int. Ed. Engl.* **37**, 868-893.
9. Wolynes, P. G., Onuchic, J. N. & Thirumalai, D. (1995). Navigating the folding routes [see comments]. *Science* **267**, 1619-20.
10. Dill, K. A. & Chan, H. S. (1997). From Levinthal to pathways to funnels. *Nature Struct.Biology.* **4**, 10-19.
11. Dinner, A. R., Sali, A., Smith, L. J., Dobson, C. M. & Karplus, M. (2000). Understanding protein folding via free-energy surfaces from theory and experiment. *Trends Biochem Sci* **25**, 331-9.
12. Jahn, T. R. & Radford, S. E. (2005). The Yin and Yang of protein folding. *Febs J* **272**, 5962-70.
13. Baldwin, R. L. (1994). Protein folding - Matching speed and stability. *Nature* **369**, 183-184.
14. Dobson, C. M. & Karplus, M. (1999). The fundamentals of protein folding: bringing together theory and experiment. *Curr Opin Struct Biol* **9**, 92-101.
15. Wolynes, P. G. (2005). Energy landscapes and solved protein-folding problems. *Philos Transact A Math Phys Eng Sci* **363**, 453-64; discussion 464-7.
16. Shea, J. E. & Brooks, C. L., 3rd. (2001). From folding theories to folding proteins: a review and assessment of simulation studies of protein folding and unfolding. *Annu Rev Phys Chem* **52**, 499-535.
17. Vendruscolo, M., Paci, E., Dobson, C. M. & Karplus, M. (2001). Three key residues form a critical contact network in a protein folding transition state. *Nature* **409**, 641-645.
18. Dobson, C. M. (1991). NMR Spectroscopy and Protein Folding - Studies of Lysozyme and alpha-Lactalbumin. *Protein Conformation.* **161:167-189**, 167-189.
19. Fersht, A. R. (2000). Transition-state structure as a unifying basis in protein-folding mechanisms: contact order, chain topology, stability, and the extended nucleus mechanism. *Proc Natl Acad Sci U S A* **97**, 1525-9.
20. Daggett, V. & Fersht, A. R. (2003). Is there a unifying mechanism for protein folding? *Trends Biochem Sci* **28**, 18-25.
21. Matouschek, A., Kellis, J. T., Serrano, L., Bycroft, M. & Fersht, A. R. (1990). Transient folding intermediates characterized by protein engineering. *Nature* **346**, 440-445.
22. Fersht, A. R. (1995). Characterizing transition states in protein folding: An essential step in the puzzle. *Curr. Opin. Struct. Biol.* **5**, 79-84.

References

23. Lindorff-Larsen, K., Rogen, P., Paci, E., Vendruscolo, M. & Dobson, C. M. (2005). Protein folding and the organization of the protein topology universe. *Trends Biochem Sci* **30**, 13-9.
24. Vendruscolo, M., Paci, E., Karplus, M. & Dobson, C. M. (2003). Structures and relative free energies of partially folded states of proteins. *Proc Natl Acad Sci U S A* **100**, 14817-21.
25. Sanchez, I. E. & Kiefhaber, T. (2003). Evidence for sequential barriers and obligatory intermediates in apparent two-state protein folding. *J. Mol. Biol.* **325**, 367-376.
26. Capaldi, A. P., Shastry, M. C., Kleanthous, C., Roder, H. & Radford, S. E. (2001). Ultrarapid mixing experiments reveal that Im7 folds via an on-pathway intermediate. *Nat. Struct. Biol.* **8**, 68-72.
27. Bollen, Y. J., Sanchez, I. E. & van Mierlo, C. P. (2004). Formation of on- and off-pathway intermediates in the folding kinetics of Azotobacter vinelandii apoflavodoxin. *Biochemistry* **43**, 10475-89.
28. Roder, H. & Colon, W. (1997). Kinetic role of early intermediates in protein folding. *Curr Opin Struct Biol* **7**, 15-28.
29. Fink, A. L. (1998). Protein aggregation: folding aggregates, inclusion bodies and amyloid. *Fold Des* **3**, R9-23.
30. Dobson, C. M. (1999). Protein misfolding, evolution and disease. *Trends Biochem. Sci.* **24**, 329-332.
31. Dobson, C. M. (2004). Principles of protein folding, misfolding and aggregation. *Semin Cell Dev Biol* **15**, 3-16.
32. Stefani, M. (2004). Protein misfolding and aggregation: new examples in medicine and biology of the dark side of the protein world. *Biochim Biophys Acta* **1739**, 5-25.
33. Dyson, H. J. & Wright, P. E. (2004). Unfolded proteins and protein folding studied by NMR. *Chem. Rev.* **104**, 3607-3622.
34. Korzhnev, D. M., Salvatella, X., Vendruscolo, M., Di Nardo, A. A., Davidson, A. R., Dobson, C. M. & Kay, L. E. (2004). Low-populated folding intermediates of Fyn SH3 characterized by relaxation dispersion NMR. *Nature* **430**, 586-90.
35. Feng, H., Zhou, Z. & Bai, Y. (2005). A protein folding pathway with multiple folding intermediates at atomic resolution. *Proc Natl Acad Sci U S A* **102**, 5026-31.
36. Ellis, R. J. (2001). Macromolecular crowding: obvious but underappreciated. *Trends Biochem Sci* **26**, 597-604.
37. Ellis, R. J. (2001). Macromolecular crowding: an important but neglected aspect of the intracellular environment. *Curr Opin Struct Biol* **11**, 114-9.
38. Bukau, B. & Horwich, A. L. (1998). The Hsp70 and Hsp60 chaperone machines. *Cell* **92**, 351-66.
39. Hartl, F. U. & Hayer-Hartl, M. (2002). Molecular chaperones in the cytosol: from nascent chain to folded protein. *Science* **295**, 1852-8.
40. Ellis, J. R. (1987). Proteins as molecular chaperones. *Nature* **328**, 378-379.
41. McClellan, A. J. & Frydman, J. (2001). Molecular chaperones and the art of recognizing a lost cause. *Nat Cell Biol* **3**, E51-3.
42. Frydman, J. (2001). Folding of newly translated proteins in vivo: the role of molecular chaperones. *Annu Rev Biochem* **70**, 603-47.
43. Mayer, M. P. & Bukau, B. (2005). Hsp70 chaperones: cellular functions and molecular mechanism. *Cell Mol Life Sci* **62**, 670-84.
44. Braig, K., Otwinowski, Z., Hegde, R., Boisvert, D. C., Joachimiak, A., Horwich, A. L. & Sigler, P. B. (1994). The crystal structure of the bacterial chaperonin GroEL at 2.8 angstrom. *Nature* **371**, 578-586.
45. Slepenkov, S. V. & Witt, S. N. (2002). The unfolding story of the Escherichia coli Hsp70 DnaK: is DnaK a holdase or an unfoldase? *Mol Microbiol* **45**, 1197-206.

46. Schiene, C. & Fischer, G. (2000). Enzymes that catalyse the restructuring of proteins. *Curr Opin Struct Biol* **10**, 40-5.
47. Fischer, G., Wittmann-Liebold, B., Lang, K., Kiefhaber, T. & Schmid, F. X. (1989). Cyclophilin and peptidyl-prolyl cis-trans isomerase are probably identical proteins. *Nature* **337**, 476-8.
48. Schmid, F. X. (1995). Prolyl isomerases join the fold. *Current Biology* **5**, 993-994.
49. Schmid, F. X. (1993). Prolyl Isomerase - Enzymatic Catalysis of Slow Protein-Folding Reactions. *Annual Review of Biophysics and Biomolecular Structure* **22**, 123-143.
50. Heras, B., Edeling, M. A., Schirra, H. J., Raina, S. & Martin, J. L. (2004). Crystal structures of the DsbG disulfide isomerase reveal an unstable disulfide. *Proc Natl Acad Sci U S A* **101**, 8876-81.
51. McCarthy, A. A., Haebel, P. W., Torronen, A., Rybin, V., Baker, E. N. & Metcalf, P. (2000). Crystal structure of the protein disulfide bond isomerase, DsbC, from Escherichia coli. *Nat Struct Biol* **7**, 196-9.
52. Saul, F. A., Arie, J. P., Vulliez-le Normand, B., Kahn, R., Betton, J. M. & Bentley, G. A. (2004). Structural and functional studies of FkpA from Escherichia coli, a cis/trans peptidyl-prolyl isomerase with chaperone activity. *J. Mol. Biol.* **335**, 595-608.
53. Hesterkamp, T. & Bukau, B. (1996). Identification of the prolyl isomerase domain of Escherichia coli trigger factor. *FEBS Lett.* **385**, 67-71.
54. Stoller, G., Rücknagel, K. P., Nierhaus, K., Schmid, F. X., Fischer, G. & Rahfeld, J.-U. (1995). Identification of the peptidyl-prolyl cis/trans isomerase bound to the Escherichia coli ribosome as the trigger factor. *EMBO Journal* **14**, 4939-4948.
55. Rouviere, P. E. & Gross, C. A. (1996). SurA, a periplasmic protein with peptidyl-prolyl isomerase activity, participates in the assembly of outer membrane porins. *Gene Develop.* **10**, 3170-3182.
56. Scholz, C., Schaarschmidt, P., Engel, A. M., Andres, H., Schmitt, U., Faatz, E., Balbach, J. & Schmid, F. X. (2004). Functional solubilization of aggregation-prone HIV envelope proteins by covalent fusion with chaperone modules. *J. Mol. Biol.*, 1229-1241.
57. Mosavi, L. K., Cammett, T. J., Desrosiers, D. C. & Peng, Z. Y. (2004). The ankyrin repeat as molecular architecture for protein recognition. *Protein Sci.* **13**, 1435-1448.
58. Marcotte, E. M., Pellegrini, M., Yeates, T. O. & Eisenberg, D. (1999). A census of protein repeats. *J Mol Biol* **293**, 151-60.
59. Groves, M. R. & Barford, D. (1999). Topological characteristics of helical repeat proteins. *Curr Opin Struct Biol* **9**, 383-9.
60. Bork, P. (1993). Hundreds of ankyrin-like repeats in functionally diverse proteins: mobile modules that cross phyla horizontally? *Proteins* **17**, 363-74.
61. Letunic, I., Goodstadt, L., Dickens, N. J., Doerks, T., Schultz, J., Mott, R., Ciccarelli, F., Copley, R. R., Ponting, C. P. & Bork, P. (2002). Recent improvements to the SMART domain-based sequence annotation resource. *Nucleic Acids Res* **30**, 242-4.
62. Main, E. R., Jackson, S. E. & Regan, L. (2003). The folding and design of repeat proteins: reaching a consensus. *Curr Opin Struct Biol* **13**, 482-9.
63. Mosavi, L. K., Minor, D. L., Jr. & Peng, Z. Y. (2002). Consensus-derived structural determinants of the ankyrin repeat motif. *Proc Natl Acad Sci U S A* **99**, 16029-34.
64. Kohl, A., Binz, H. K., Forrer, P., Stumpp, M. T., Plückthun, A. & Grütter, M. G. (2003). Designed to be stable: crystal structure of a consensus ankyrin repeat protein. *Proc Natl Acad Sci U S A* **100**, 1700-5.
65. Binz, H. K., Amstutz, P., Kohl, A., Stumpp, M. T., Briand, C., Forrer, P., Grütter, M. G. & Plückthun, A. (2004). High-affinity binders selected from designed ankyrin repeat protein libraries. *Nat. Biotechnol.* **22**, 575-582.

References

66. Amstutz, P., Binz, H. K., Parizek, P., Stumpp, M. T., Kohl, A., Grutter, M. G., Forrer, P. & Pluckthun, A. (2005). Intracellular kinase inhibitors selected from combinatorial libraries of designed ankyrin repeat proteins. *J Biol Chem* **280**, 24715-22.
67. Kohl, A., Amstutz, P., Parizek, P., Binz, H. K., Briand, C., Capitani, G., Forrer, P., Pluckthun, A. & Grutter, M. G. (2005). Allosteric inhibition of aminoglycoside phosphotransferase by a designed ankyrin repeat protein. *Structure* **13**, 1131-41.
68. Amstutz, P., Koch, H., Binz, H. K., Deuber, S. A. & Pluckthun, A. (2006). Rapid selection of specific MAP kinase-binders from designed ankyrin repeat protein libraries. *Protein Eng Des Sel* **19**, 219-29.
69. Zahnd, C., Wyler, E., Schwenk, J. M., Steiner, D., Lawrence, M. C., McKern, N. M., Pecorari, F., Ward, C. W., Joos, T. O. & Pluckthun, A. (2007). A designed ankyrin repeat protein evolved to picomolar affinity to Her2. *J Mol Biol* **369**, 1015-28.
70. Schweizer, A., Roschitzki-Voser, H., Amstutz, P., Briand, C., Gulotti-Georgieva, M., Prenosil, E., Binz, H. K., Capitani, G., Baici, A., Pluckthun, A. & Grutter, M. G. (2007). Inhibition of caspase-2 by a designed ankyrin repeat protein: specificity, structure, and inhibition mechanism. *Structure* **15**, 625-36.
71. Main, E. R., Lowe, A. R., Mochrie, S. G., Jackson, S. E. & Regan, L. (2005). A recurring theme in protein engineering: the design, stability and folding of repeat proteins. *Curr Opin Struct Biol* **15**, 464-71.
72. Kloss, E., Courtemanche, N. & Barrick, D. (2008). Repeat-protein folding: new insights into origins of cooperativity, stability, and topology. *Arch Biochem Biophys* **469**, 83-99.
73. Sedgwick, S. G. & Smerdon, S. J. (1999). The ankyrin repeat: a diversity of interactions on a common structural framework. *Trends Biochem Sci* **24**, 311-6.
74. Michaely, P., Tomchick, D. R., Machius, M. & Anderson, R. G. (2002). Crystal structure of a 12 ANK repeat stack from human ankyrinR. *Embo J* **21**, 6387-96.
75. Breeden, L. & Nasmyth, K. (1987). Similarity between cell-cycle genes of budding yeast and fission yeast and the Notch gene of Drosophila. *Nature* **329**, 651-4.
76. Lux, S. E., John, K. M. & Bennett, V. (1990). Analysis of cDNA for human erythrocyte ankyrin indicates a repeated structure with homology to tissue-differentiation and cell-cycle control proteins. *Nature* **344**, 36-42.
77. Zhang, B. & Peng, Z. (2000). A minimum folding unit in the ankyrin repeat protein p16(INK4). *J Mol Biol* **299**, 1121-1132.
78. Morgan, D. O. (1995). Principles of CDK regulation. *Nature* **374**, 131-134.
79. Sherr, C. J. & Roberts, J. M. (1999). CDK inhibitors: positive and negative regulators of G1-phase progression. *Genes. Dev.* **13**, 1501-1512.
80. Sherr, C. J. (1996). Cancer cell cycles. *Science* **274**, 1672-1677.
81. Ren, G., Vajjhala, P., Lee, J. S., Winsor, B. & Munn, A. L. (2006). The BAR domain proteins: molding membranes in fission, fusion, and phagy. *Microbiol. Mol. Biol. Rev.* **70**, 37-120.
82. McMahon, H. T. & Gallop, J. L. (2005). Membrane curvature and mechanisms of dynamic cell membrane remodelling. *Nature* **438**, 590-6.
83. David, C., McPherson, P. S., Mundigl, O. & de Camilli, P. (1996). A role of amphiphysin in synaptic vesicle endocytosis suggested by its binding to dynamin in nerve terminals. *Proc. Natl. Acad. Sci. U S A* **93**, 331-5.
84. Zhang, B. & Zelhof, A. C. (2002). Amphiphysins: raising the BAR for synaptic vesicle recycling and membrane dynamics. Bin-Amphiphysin-Rvsp. *Traffic* **3**, 452-60.
85. Dawson, J. C., Legg, J. A. & Machesky, L. M. (2006). Bar domain proteins: a role in tubulation, scission and actin assembly in clathrin-mediated endocytosis. *Trends Cell Biol.* **16**, 493-8.

86. Henne, W. M., Kent, H. M., Ford, M. G., Hegde, B. G., Daumke, O., Butler, P. J., Mittal, R., Langen, R., Evans, P. R. & McMahon, H. T. (2007). Structure and analysis of FCHo2 F-BAR domain: a dimerizing and membrane recruitment module that effects membrane curvature. *Structure* **15**, 839-52.
87. Peter, B. J., Kent, H. M., Mills, I. G., Vallis, Y., Butler, P. J., Evans, P. R. & McMahon, H. T. (2004). BAR domains as sensors of membrane curvature: the amphiphysin BAR structure. *Science* **303**, 495-9.
88. Gallop, J. L., Jao, C. C., Kent, H. M., Butler, P. J., Evans, P. R., Langen, R. & McMahon, H. T. (2006). Mechanism of endophilin N-BAR domain-mediated membrane curvature. *EMBO J.* **25**, 2898-910.
89. Casal, E., Federici, L., Zhang, W., Fernandez-Recio, J., Priego, E. M., Miguel, R. N., DuHadaway, J. B., Prendergast, G. C., Luisi, B. F. & Laue, E. D. (2006). The crystal structure of the BAR domain from human Bin1/amphiphysin II and its implications for molecular recognition. *Biochemistry* **45**, 12917-28.
90. Zimmerberg, J. & Kozlov, M. M. (2006). How proteins produce cellular membrane curvature. *Nat. Rev. Mol. Cell Biol.* **7**, 9-19.
91. Reynwar, B. J., Illya, G., Harmandaris, V. A., Muller, M. M., Kremer, K. & Deserno, M. (2007). Aggregation and vesiculation of membrane proteins by curvature-mediated interactions. *Nature* **447**, 461-4.
92. Frolov, V. A. & Zimmerberg, J. (2008). Flexible scaffolding made of rigid BARs. *Cell* **132**, 727-9.
93. Shnyrova, A. V., Ayllon, J., Mikhalyov, II, Villar, E., Zimmerberg, J. & Frolov, V. A. (2007). Vesicle formation by self-assembly of membrane-bound matrix proteins into a fluidlike budding domain. *J Cell Biol* **179**, 627-33.
94. Masuda, M., Takeda, S., Sone, M., Ohki, T., Mori, H., Kamioka, Y. & Mochizuki, N. (2006). Endophilin BAR domain drives membrane curvature by two newly identified structure-based mechanisms. *EMBO J.* **25**, 2889-97.
95. Blood, P. D. & Voth, G. A. (2006). Direct observation of Bin/amphiphysin/Rvs (BAR) domain-induced membrane curvature by means of molecular dynamics simulations. *Proc. Natl. Acad. Sci. U S A* **103**, 15068-72.
96. Morth, J. P., Pedersen, B. P., Toustrup-Jensen, M. S., Sorensen, T. L., Petersen, J., Andersen, J. P., Vilsen, B. & Nissen, P. (2007). Crystal structure of the sodium-potassium pump. *Nature* **450**, 1043-9.
97. Cohen, S. N., Chang, A. C., Boyer, H. W. & Helling, R. B. (1973). Construction of biologically functional bacterial plasmids in vitro. *Proc Natl Acad Sci U S A* **70**, 3240-4.
98. Johnson, I. S. (1983). Human insulin from recombinant DNA technology. *Science* **219**, 632-7.
99. Ventura, S. & Villaverde, A. (2006). Protein quality in bacterial inclusion bodies. *Trends Biotechnol* **24**, 179-85.
100. Piatak, M., Lane, J. A., Laird, W., Bjorn, M. J., Wang, A. & Williams, M. (1988). Expression of soluble and fully functional ricin A chain in Escherichia coli is temperature-sensitive. *J Biol Chem* **263**, 4837-43.
101. Turner, P., Holst, O. & Karlsson, E. N. (2005). Optimized expression of soluble cyclomaltodextrinase of thermophilic origin in Escherichia coli by using a soluble fusion-tag and by tuning of inducer concentration. *Protein Expr Purif* **39**, 54-60.
102. Covalt, J. C., Jr., Cao, T. B., Magdaroag, J. R., Gross, L. A. & Jennings, P. A. (2005). Temperature, media, and point of induction affect the N-terminal processing of interleukin-1beta. *Protein Expr Purif* **41**, 45-52.
103. Brotherton, D. H., Dhanaraj, V., Wick, S., Brizuela, L., Domaille, P. J., Volyanik, E., Xu, X., Parisini, E., Smith, B. O., Archer, S. J., Serrano, M., Brenner, S. L., Blundell,

T. L. & Laue, E. D. (1998). Crystal structure of the complex of the cyclin D-dependent kinase Cdk6 bound to the cell-cycle inhibitor p19INK4d. *Nature* **395**, 244-250.

104. Baumgartner, R., Fernandez-Catalan, C., Winoto, A., Huber, R., Engh, R. A. & Holak, T. A. (1998). Structure of human cyclin-dependent kinase inhibitor p19INK4d: comparison to known ankyrin-repeat-containing structures and implications for the dysfunction of tumor suppressor p16INK4a. *Structure* **6**, 1279-1290.

105. Mosavi, L. K., Williams, S. & Peng, Z. Y. (2002). Equilibrium folding and stability of myotrophin: a model ankyrin repeat protein. *J. Mol. Biol.* **320**, 165-170.

106. Tang, K. S., Guralnick, B. J., Wang, W. K., Fersht, A. R. & Itzhaki, L. S. (1999). Stability and folding of the tumour suppressor protein p16. *J. Mol. Biol.* **285**, 1869-1886.

107. Zweifel, M. E. & Barrick, D. (2001). Studies of the ankyrin repeats of the Drosophila melanogaster Notch receptor. 2. Solution stability and cooperativity of unfolding. *Biochemistry* **40**, 14357-14367.

108. Religa, T. L., Markson, J. S., Mayor, U., Freund, S. M. & Fersht, A. R. (2005). Solution structure of a protein denatured state and folding intermediate. *Nature* **437**, 1053-6.

109. Lowe, A. R. & Itzhaki, L. S. (2007). Rational redesign of the folding pathway of a modular protein. *Proc. Natl. Acad. Sci. USA* **104**, 2679-2684.

110. Werbeck, N. D. & Itzhaki, L. S. (2007). Probing a moving target with a plastic unfolding intermediate of an ankyrin-repeat protein. *Proc. Natl. Acad. Sci. USA* **104**, 7863-7868.

111. Thullberg, M., Bartkova, J., Khan, S., Hansen, K., Ronnstrand, L., Lukas, J., Strauss, M. & Bartek, J. (2000). Distinct versus redundant properties among members of the INK4 family of cyclin-dependent kinase inhibitors. *FEBS Lett.* **470**, 161-166.

112. Scassa, M. E., Marazita, M. C., Ceruti, J. M., Carcagno, A. L., Sirkin, P. F., Gonzzalez-Cid, M., Pignataro, O. P. & Canepa, E. T. (2007). Cell cycle inhibitor, p19INK4d, promotes cell survival and decreasese chromosomal aberrations after genotoxic insult due to enhanced DNA repair. *DNA repair*, [Epub ahead of print].

113. Ceruti, J. M., Scassa, M. E., Flo, J. M., Varone, C. L. & Canepa, E. T. (2005). Induction of p19INK4d in response to ultraviolet light improves DNA repair and confers resistance to apoptosis in neuroblastoma cells. *Oncogene* **24**, 4065-4080.

114. Thullberg, M., Bartek, J. & Lukas, J. (2000). Ubiquitin/proteasome-mediated degradation of p19INK4d determines its periodic expression during the cell cycle. *Oncogene* **19**, 2870-2876.

115. Hupp, T. R. & Lane, D. P. (1995). Two distinct signaling pathways activate the latent DNA binding function of p53 in a casein kinase II-independent manner. *J Biol Chem* **270**, 18165-74.

116. Park, K. S., Mohapatra, D. P., Misonou, H. & Trimmer, J. S. (2006). Graded regulation of the Kv2.1 potassium channel by variable phosphorylation. *Science* **313**, 976-9.

117. Koepp, D. M., Schaefer, L. K., Ye, X., Keyomarsi, K., Chu, C., Harper, J. W. & Elledge, S. J. (2001). Phosphorylation-dependent ubiquitination of cyclin E by the SCFFbw7 ubiquitin ligase. *Science* **294**, 173-7.

118. Hao, B., Oehlmann, S., Sowa, M. E., Harper, J. W. & Pavletich, N. P. (2007). Structure of a Fbw7-Skp1-cyclin E complex: multisite-phosphorylated substrate recognition by SCF ubiquitin ligases. *Mol Cell* **26**, 131-43.

119. Kawashima, T., Amano, N., Koike, H., Makino, S., Higuchi, S., Kawashima-Ohya, Y., Watanabe, K., Yamazaki, M., Kanehori, K., Kawamoto, T., Nunoshiba, T., Yamamoto, Y., Aramaki, H., Makino, K. & Suzuki, M. (2000). Archaeal adaptation to

higher temperatures revealed by genomic sequence of Thermoplasma volcanium. *Proc Natl Acad Sci U S A* **97**, 14257-62.

120. Watters, A. L., Deka, P., Corrent, C., Callender, D., Varani, G., Sosnick, T. & Baker, D. (2007). The highly cooperative folding of small naturally occurring proteins is likely the result of natural selection. *Cell* **128**, 613-24.
121. Truhlar, S. M., Torpey, J. W. & Komives, E. A. (2006). Regions of IkappaBalpha that are critical for its inhibition of NF-kappaB.DNA interaction fold upon binding to NF-kappaB. *Proc Natl Acad Sci U S A* **103**, 18951-6.
122. Kloss, E., Courtemanche, N. & Barrick, D. (2007). Repeat-protein folding: New insights into origins of cooperativity, stability, and topology. *Arch Biochem Biophys*, in press.
123. Mello, C. C. & Barrick, D. (2004). An experimentally determined protein folding energy landscape. *Proc Natl Acad Sci U S A* **101**, 14102-14107.
124. Street, T. O., Bradley, C. M. & Barrick, D. (2007). Predicting coupling limits from an experimentally determined energy landscape. *Proc Natl Acad Sci U S A* **104**, 4907-4912.
125. Kajander, T., Cortajarena, A. L., Main, E. R., Mochrie, S. G. & Regan, L. (2005). A new folding paradigm for repeat proteins. *J Am Chem Soc* **127**, 10188-90.
126. Scholz, C., Eckert, B., Hagn, F., Schaarschmidt, P., Balbach, J. & Schmid, F. X. (2006). SlyD proteins from different species exhibit high prolyl isomerase and chaperone activities. *Biochemistry* **45**, 20-33.
127. Bernhardt, T. G., Roof, W. D. & Young, R. (2002). The Escherichia coli FKBP-type PPIase SlyD is required for the stabilization of the E lysis protein of bacteriophage phi X174. *Mol. Microbiol.* **45**, 99-108.
128. Roof, W. D., Fang, H. Q., Young, K. D., Sun, J. & Young, R. (1997). Mutational analysis of slyD, an Escherichia coli gene encoding a protein of the FKBP immunophilin family. *Mol Microbiol* **25**, 1031-46.
129. Roof, W. D., Horne, S. M., Young, K. D. & Young, R. (1994). slyD, a host gene required for phi X174 lysis, is related to the FK506-binding protein family of peptidyl-prolyl cis-trans-isomerases. *J Biol Chem* **269**, 2902-10.
130. Roof, W. D. & Young, R. (1995). Phi X174 lysis requires slyD, a host gene which is related to the FKBP family of peptidyl-prolyl cis-trans isomerases. *FEMS Microbiol Rev* **17**, 213-8.
131. Weiwad, M., Werner, A., Rucknagel, P., Schierhorn, A., Kullertz, G. & Fischer, G. (2004). Catalysis of proline-directed protein phosphorylation by peptidyl-prolyl cis/trans isomerases. *J Mol Biol* **339**, 635-46.
132. Schreiber, S. L. & Crabtree, G. R. (1992). The Mechanism of Action of Cyclosporin-A and FK506. *Immunol.Today* **13**, 136-142.
133. Scholz, C., Thirault, L., Schaarschmidt, P., Zarnt, T., Faatz, E., Engel, A. M., Upmeier, B., Bollhagen, R., Eckert, B. & Schmid, F. X. (2008). Chaperone-Aided in Vitro Renaturation of an Engineered E1 Envelope Protein for Detection of Anti-Rubella Virus IgG Antibodies. *Biochemistry*.
134. Henzler-Wildman, K. & Kern, D. (2007). Dynamic personalities of proteins. *Nature* **450**, 964-72.
135. Henzler-Wildman, K. A., Lei, M., Thai, V., Kerns, S. J., Karplus, M. & Kern, D. (2007). A hierarchy of timescales in protein dynamics is linked to enzyme catalysis. *Nature* **450**, 913-6.
136. Henzler-Wildman, K. A., Thai, V., Lei, M., Ott, M., Wolf-Watz, M., Fenn, T., Pozharski, E., Wilson, M. A., Petsko, G. A., Karplus, M., Hubner, C. G. & Kern, D. (2007). Intrinsic motions along an enzymatic reaction trajectory. *Nature* **450**, 838-44.

137. Shupliakov, O., Low, P., Grabs, D., Gad, H., Chen, H., David, C., Takei, K., De Camilli, P. & Brodin, L. (1997). Synaptic vesicle endocytosis impaired by disruption of dynamin-SH3 domain interactions. *Science* **276**, 259-63.
138. Lee, E., Marcucci, M., Daniell, L., Pypaert, M., Weisz, O. A., Ochoa, G. C., Farsad, K., Wenk, M. R. & De Camilli, P. (2002). Amphiphysin 2 (Bin1) and T-tubule biogenesis in muscle. *Science* **297**, 1193-6.
139. Itoh, T. & De Camilli, P. (2006). BAR, F-BAR (EFC) and ENTH/ANTH domains in the regulation of membrane-cytosol interfaces and membrane curvature. *Biochim Biophys Acta* **1761**, 897-912.

5. List of Publications

A) Christian Löw, Ulrich Weininger, Markus Zeeb, Wei Zhang, Ernest D. Laue, Franz X. Schmid and Jochen Balbach.
Folding Mechanism of an Ankyrin Repeat Protein: Scaffold and Active Site Formation of Human CDK Inhibitor p19^{INK4d}.
Journal of Molecular Biology **373**, 219-231 (2007)
(Faculty of 1000 - evaluation)

B) Christian Löw, Nadine Homeyer, Ulrich Weininger, Heinrich Sticht and Jochen Balbach.
Conformational Switch upon Phosphorylation: Human CDK Inhibitor p19^{INK4d} Between Native and Partially Folded State.
(submitted manuscript)

C) Christian Löw, Ulrich Weininger, Piotr Neumann, Mirjam Klepsch, Hauke Lilie, Milton T. Stubbs and Jochen Balbach.
Structural Insights into an Equilibrium Folding Intermediate of an Archaeal Ankyrin Repeat Protein.
Proc. Natl. Acad. Sci. **105**, 3779-3784 (2008)

D) Christian Löw, Piotr Neumann, Henning Tidow, Ulrich Weininger, Beatrice Epler, Christian Scholz, Milton T. Stubbs and Jochen Balbach.
Crystal Structure and Functional Characterization of the Thermophilic Prolyl Isomerase and Chaperone SlyD.
(prepared manuscript)

E) Christian Löw, Ulrich Weininger, Hwankyu Lee, Kristian Schweimer, Ines Neundorf, Annette G. Beck-Sickinger, Richard W. Pastor and Jochen Balbach.
Structure and Dynamics of Helix-0 of the N-BAR Domain in Lipid Micelles and Bilayers.
(*Biophysical Journal*, revision)

Additional publications:

F) Markus Zeeb, Klaas E.A. Max, Ulrich Weininger, Christian Löw, Heinrich Sticht and Jochen Balbach.
Recognition of T-rich single-stranded DNA by the Cold Shock Protein BS-CspB in Solution.
Nucl. Acids Res. **34**, 4561-4571 (2006)

G) Cindy Schulenburg, Maria M. Martinez-Senac, Christian Löw, Ralph Golbik, Renate Ulbrich-Hofmann and Ulrich Arnold.
Identifiaction of three Phases in Onconase Refolding.
FEBS Journal **274**, 5826-5833 (2007)

6. Presentation of personal contribution

I designed and performed the experiments on the following projects. Collaborations and contributions of coworkers are stated.

A) The publication describes the folding mechanism of the AR protein p19. Markus Zeeb introduced me into the basic methods of protein folding. NMR H/D exchange spectroscopy was done together with Ulrich Weininger. In the lab of Ernest D. Laue, I expressed human CDK6 in the baculovirus/insect cell system under supervision of Wei Zhang. Franz X. Schmid, Jochen Balbach and I wrote the publication.

B) The manuscript describes the influence of phosphorylation on stability, folding and function of the AR protein p19. NMR spectroscopy was done together with Ulrich Weininger. Nadine Homeyer in the lab of Heinrich Sticht performed and analysed MD simulations. Heinrich Sticht and I wrote the manuscript.

C) The publication describes the first structure of a thermophilic AR protein including a detailed analysis of the folding mechanism. Backbone assignment was done by Ulrich Weininger. NMR GdmCl equilibrium analysis was performed together with Ulrich Weininger. X-ray data were analyzed by Piotr Neumann in the lab of Milton Stubbs. Hauke Lilie performed ultracentrifugation analyses and Mirjam Klepsch mass spectroscopy. Jochen Balbach and I wrote the publication.

D) The manuscript presents a structural and functional analysis of the thermophilic prolyl isomerase and chaperone SlyD. Backbone assignment was done by Ulrich Weininger. NMR H/D exchange was performed together with Ulrich Weininger. Beatrice Epler helped with crystallization. X-ray data were analyzed by Piotr Neumann in the lab of Milton Stubbs. Henning Tidow performed SAXS measurement and analyzed data. Christian Scholz gave useful advice throughout the entire project. I wrote the manuscript.

E) The manuscript describes the structure and functional importance of the amphipathic helix of BAR domains. Kristian Schweimer measured NMR assignment spectra. Ulrich Weininger calculated NMR structures and performed NMR dynamic studies. Ines Neundorf in the lab of Annette G. Beck-Sickinger synthesized the peptide for initial experiments. Hwankyu Lee in the lab of Richard W. Pastor performed and analysed MD simulations. Richard W. Pastor, Jochen Balbach and I wrote the manuscript.

7. Subprojects

7.1 Subproject A

Subproject A

Subproject A

doi:10.1016/j.jmb.2007.07.063

Available online at www.sciencedirect.com
ScienceDirect

J. Mol. Biol. (2007) 373, 219–231

Folding Mechanism of an Ankyrin Repeat Protein: Scaffold and Active Site Formation of Human CDK Inhibitor p19^{INK4d}

Christian Löw[1], Ulrich Weininger[1], Markus Zeeb[2], Wei Zhang[3]
Ernest D. Laue[3], Franz X. Schmid[2] and Jochen Balbach[1]*

[1]*Institut für Physik, Biophysik und Mitteldeutsches Zentrum für Struktur und Dynamik der Proteine (MZP) Martin-Luther-Universität Halle-Wittenberg, D-06120 Halle(Saale), Germany*

[2]*Laboratorium für Biochemie Universität Bayreuth D-95440 Bayreuth, Germany*

[3]*Department of Biochemistry University of Cambridge 80 Tennis Court Road Cambridge CB2 1GA, UK*

Received 2 June 2007;
received in revised form
22 July 2007;
accepted 26 July 2007
Available online
10 August 2007

Edited by C. R. Matthews

The p19^{INK4d} protein consists of five ankyrin repeats (ANK) and controls the human cell cycle by inhibiting the cyclin D-dependent kinases (CDK) 4 and 6. We investigated the folding of p19^{INK4d} by urea-induced unfolding transitions, kinetic analyses of unfolding and refolding, including double-mixing experiments and a special assay for folding intermediates. Folding is a sequential two-step reaction *via* a hyperfluorescent on-pathway intermediate. This intermediate is present under all conditions, during unfolding, refolding and at equilibrium. The folding mechanism was confirmed by a quantitative global fit of a consistent set of equilibrium and kinetic data revealing the thermodynamics and intrinsic folding rates of the different states. Surprisingly, the N↔I transition is much faster compared to the I↔U transition. The urea-dependence of the intrinsic folding rates causes population of the intermediate at equilibrium close to the transition midpoint. NMR detected hydrogen/deuterium exchange and the analysis of truncated variants showed that the C-terminal repeats ANK3-5 are already folded in the on-pathway intermediate, whereas the N-terminal repeats 1 and 2 are not folded. We suggest that during refolding, repeats ANK3–ANK5 first form the scaffold for the subsequent assembly of repeats ANK1 and ANK2. The binding function of p19^{INK4d} resides in the latter repeats. We propose that the graded stability and the facile unfolding of repeats 1 and 2 is a prerequisite for the down-regulation of the inhibitory activity of p19^{INK4d} during the cell-cycle.

© 2007 Elsevier Ltd. All rights reserved.

Keywords: ankyrin repeat; protein folding; p19^{INK4d}; folding intermediate; folding kinetics

Introduction

The ankyrin repeat (ANK) is one of the commonest structural motifs found in proteins. About 20,000 repeats have been identified in more than 3500 proteins.[1] Their importance is underlined by their abundance across all kingdoms of life, and their function in numerous fundamental physiological processes. As in other repeat protein families, such as leucine-rich or armadillo repeats, ANKs form the scaffold for specific, high-affinity molecular interactions. Each repeat consists of 33 amino acid residues comprising a loop connecting the preceding ANK and a β-turn followed by two antiparallel α-helices. Usually, only four to six repeats stack onto each other to form an elongated structure with a continuous hydrophobic core and a large solvent-accessible surface but up to 29 repeats have been found in a single protein.[1]

The INK4 proteins are also composed of ankyrin repeats. Their four members (p16^{INK4a}, p15^{INK4b}, p18^{INK4c}, and p19^{INK4d}) negatively regulate the mammalian cell-cycle by specific inhibition of the

*Corresponding author. E-mail address:
jochen.balbach@physik.uni-halle.de.
Present address: M. Zeeb, Department of Lead Discovery, Boehringer Ingelheim Pharma GmbH and Co. KG, D-88397 Biberach an der Riss, Germany.
Abbreviations used: ANK, ankyrin repeat; CDK, cyclin D-dependent kinase; H/D, hydrogen/deuterium; HSQC, heteronuclear single quantum coherence.

0022-2836/$ - see front matter © 2007 Elsevier Ltd. All rights reserved.

two cyclin D-dependent kinases (CDK) CDK4 and CDK6. The latter are involved in phosphorylation of the retinoblastoma tumor-suppressor protein and thereby in controlling the G1 to S transition in the cell-cycle.[2] Proteins of the INK4 family are hot spots for mutations in many types of tumors, such as familial melanoma or bladder cancer.[3] INK4 proteins are structurally homologous and harbor redundant as well as non-overlapping properties, including phosphorylation,[4] degradation[5] and genomic stability under genotoxic stress.[6,7]

There is a vital interest in studying the protein folding of ANK proteins.[8–16] Because of the low contact order (i.e. the average sequence separation between contacting residues) of the modular architecture very high refold rates are expected.[17] Experimentally, however, most ANK proteins were found to fold much slower than predicted because of the slow formation of protein folding intermediates.[9] A detailed analysis of the folding mechanism of ANK proteins including p16[INK4a],[18] the ankyrin domain of the *Drosophila* Notch receptor (Notch ankyrin domain),[14] and myotrophin[10,19] is difficult, because the wild-type proteins typically show two-state unfolding transitions at equilibrium but complex folding kinetics. Therefore, extended mutational studies have been performed, to elucidate indirectly the folding and local stability of individual ankyrin repeats.[11–13] Knowledge about folding and the role of single ankyrin repeats is important for designing very stable and highly active ANK proteins, which has already been successful for pharmaceutical and biotechnological applications.[20–22]

Here, we provide a detailed biophysical analysis of the folding mechanism of p19[INK4d] consisting of five sequentially arranged ankyrin repeats (Figure 1).[23] Double-mixing experiments revealed a kinetic intermediate during both unfolding and refolding. A global analysis of CD and fluorescence detected equilibrium folding transitions and the complex unfolding and refolding kinetics of p19[INK4d] confirmed a sequential folding pathway including a hyperfluorescent intermediate. The same intermediate state is populated up to 15% even at equilibrium. NMR experiments with p19[INK4d] and kinetic data for a truncated variant suggest that the C-terminal ankyrin repeats 3–5 of p19[INK4d] fold first and provide a scaffold for the subsequent folding of the N-terminal repeats 1 and 2, which exert the inhibitory function against CDK4/6.

Results

Functional analysis of p19[INK4d] wild-type and variants

Wild-type p19[INK4d] is devoid of tyrosine or tryptophan residues, and thus it lacks sensitive fluorescent probes for the analysis of the folding mechanism by fluorescence spectroscopy. To introduce a Trp residue into the central repeat (ANK3), we generated the variants p19 F86W and p19 H96W

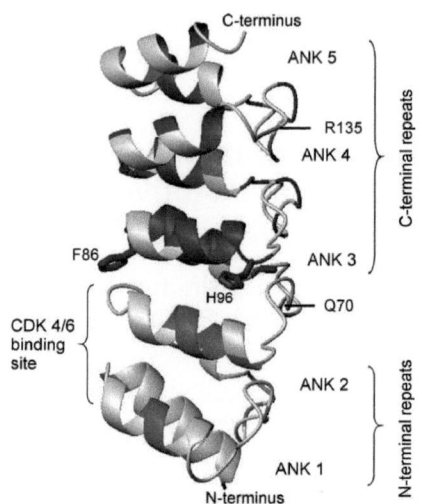

Figure 1. MOLMOL[45] representation of the crystal structure of p19[INK4d] (1bd8.pdb from the Protein Data Base).[46] Five ankyrin repeats (ANK1–ANK5), each comprising a loop, a β-turn and two sequential α-helices form the elongated structure, where Phe86 and His96 are indicated by a stick illustration of the side-chains. Residues with highly protected backbone amide protons against solvent exchange ($P > 12,000$) are indicated in blue and less protected backbone amides ($P < 12,000$) in red. Q70 denotes the first residue of truncated p19 ANK3–5 containing C-terminal ankyrin repeat 3, 4, and 5 of the wild-type protein and R135 denotes the last residue of truncated p19 ANK1–4. The CDK4/6 binding site is formed mainly by the N-terminal ankyrin repeats 1 and 2.

(Figure 1). Both protein variants with the single Trp mutations were functional and bound to human cyclin D-dependent kinase 6 (CDK6), which is the natural substrate of the p19[INK4d] inhibitor (Figure 2).[2] Several other positions for placing a Trp residue were tested. Trp86 was the best reporter; it did not affect the function, and it provided the most sensitive probe to study the kinetic folding intermediate. Therefore, the F86W variant was used for most folding experiments (see below). In addition to the full-length protein, a truncated variant of p19[INK4d] has been produced, comprising only the C-terminal ankyrin repeats 3, 4, and 5, and Trp at position 86 (p19 ANK3-5). The crystal structure of human CDK6 in complex with p16[INK4a] and p19[INK4d] revealed that binding to the kinase is mediated mainly by ankyrin repeats 1 and 2 (Figure 1).[23,24] Accordingly, truncation of ankyrin repeats 1 and 2 (p19 ANK3–5 in Figure 2) abolished CDK binding. The homolog p16[INK4a] comprises only four ankyrin repeats.[24] The respectively truncated p19 ANK1–4 variant was still binding CDK6, indicating that ANK5 is not necessary for complex formation.

Folding Mechanism of Ankyrin Repeat Protein p19INK4d

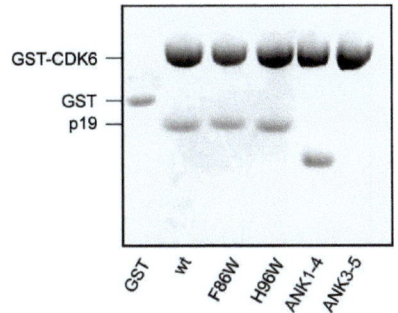

Figure 2. SDS-PAGE analysis of the pull-down assay of wild-type p19[INK4d] and different variants. Immobilized GST-CDK6 on glutathione Sepharose bound wild-type p19[INK4d], p19 F86W, p19 H96W, and p19 ANK1–4, but not p19 ANK3–5.

Stabilities of p19[INK4d] variants

The F86W mutation left the thermodynamic stability of p19[INK4d] unchanged, and virtually identical urea-induced unfolding transitions were observed for the wild-type protein[8] and the F86W variant when monitored either by CD or by fluorescence (Table 1). All transition midpoints analyzed by a two-state model were close to 3 M urea. Truncation of repeats from either end of the molecule led to destabilization. For p19 ANK3–5, the midpoint was reduced to 2.1 M urea and for p19 ANK1–4 to 1.6 M urea (see Table 1). As expected from designed ankyrin repeat protein libraries,[20] p19[INK4d] also requires an optimized C-terminal repeat to achieve a high levell of thermodynamic stability. Truncation of ankyrin repeat 5 resulted in a loss of stability, which is equivalent to the difference in stability between p19[INK4d] and the four ankyrin repeats comprising p16[INK4a]. This finding confirms the earlier conclusion that p19[INK4d] containing a fifth, stabilizing ankyrin repeat might complement the function of less stable INK4 inhibitors in cell-cycle control under unfavorable conditions.[8] The wild-type protein and the truncated forms showed identical far UV-CD spectra (when plotted as mean residue weight ellipticity), indicating that the remaining ankyrin repeats in the truncated variants contain native-like α-helical structure.

Folding kinetics of p19 F86W and truncated p19 ANK3–5

Single-mixing unfolding and refolding kinetics of p19 F86W were measured by stopped-flow fluorescence spectroscopy. Unfolding is a biphasic process (see overshoot kinetics in Figure 3(a)). The fluorescence of Trp86 increases strongly during the first second of unfolding, reaches a maximum at about 1.3 s and then decreases strongly to a final value that is lower than the fluorescence of the folded protein. This indicates that an intermediate becomes populated transiently during unfolding with strongly increased fluorescence of Trp86 relative to both the native and the unfolded state. The refolding reaction of unfolded p19 F86W is also biphasic and the fluorescence increases in both phases (Figure 3(b)). Unfolding and refolding were also measured by stopped-flow far UV-CD spectroscopy (Figure 3(c) and (d), respectively). The unfolding kinetics were biphasic as well, and the rate constants coincided with those measured by Trp fluorescence (see below). The unfolding kinetics showed a lag, which supports the conclusion from the fluorescence-detected unfolding kinetics that unfolding is a sequential process. Far UV-CD detected refolding kinetics were monophasic (Figure 3(d)) and corresponded very well to the slow refolding phase detected by fluorescence. Unfolding and refolding rates of wild-type p19[INK4d] was determined earlier by using a stopped-flow CD device with a lower

Table 1. Thermodynamic data of urea-induced unfolding of different p19[INK4d] variants at 15 °C

Method	p19[INK4d] variant	ΔG^0_{NU} (kJ/mol)	ΔG^0_{NI} (kJ/mol)	ΔG^0_{IU} (kJ/mol)	m_{NU} (kJ mol^{-1} M^{-1})	m_{NI} (kJ mol^{-1} M^{-1})	m_{IU} (kJ mol^{-1} M^{-1})
Equilibrium CD	wt	27.3			8.88		
Equilibrium fluorescence	wt	27.8			9.11		
Equilibrium CD	F86W	26.4			8.64		
Equilibrium fluorescence	F86W	26.5			8.48		
Kinetics	F86W	26.4	14.8	12.3	9.26[a]	4.60[b]	4.66[c]
I assay	F86W	30.6[d]	19.1[d]	11.5	10.2[a]	5.5[b]	4.7[c]
Equilibrium CD	ANK1–4[e]	12.4			7.62		
Equilibrium CD	ANK3–5	9.23			4.8		
Equilibrium fluorescence	ANK3–5	9.62			5.03		
Kinetics	ANK3–5	7.97			4.13		

[a] $m_{NU} = m_{NI} + m_{IU}$.
[b] $m_{NI} = m^{kin}_{NI} - m^{kin}_{IN}$.
[c] $m_{IU} = m^{kin}_{IU} - m^{kin}_{UI}$.
[d] ΔG^0_{NU} and ΔG^0_{NI} were overestimated by about 4 kJ/mol, probably because the unknown urea-dependence of the I fluorescence had been assumed to coincide with N.
[e] In the p19 ANK1–4 variant, H96 was additionally substituted by tryptophan.

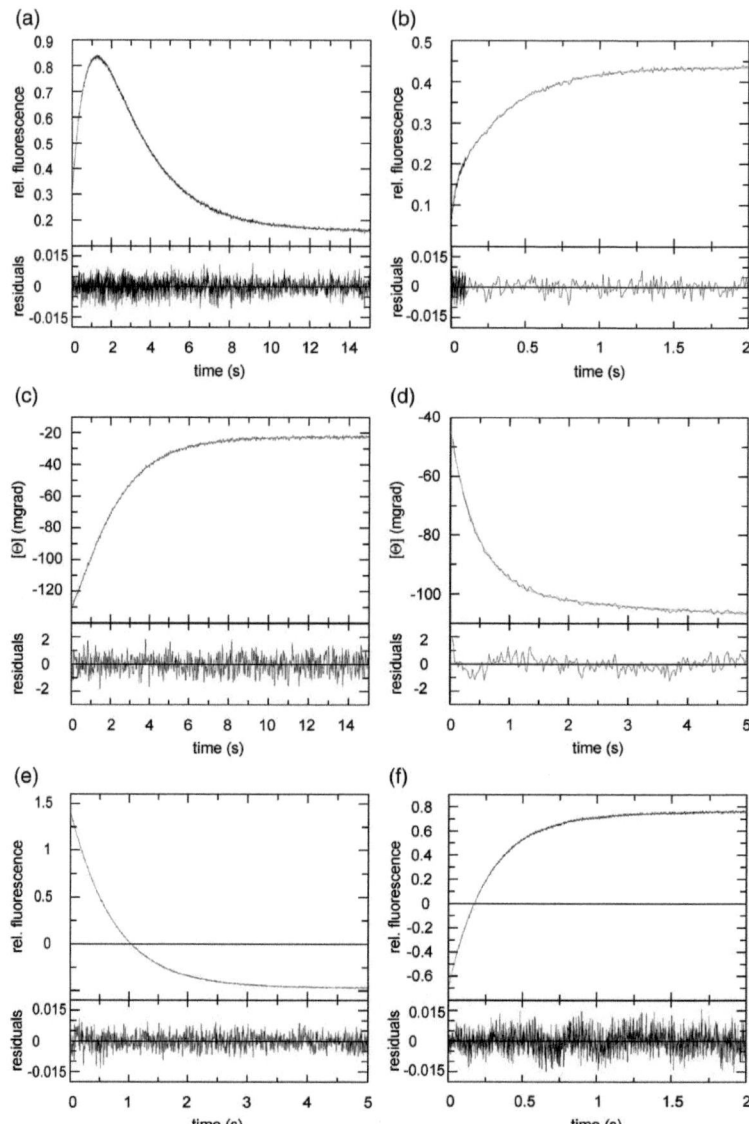

Figure 3. Single-mixing unfolding and refolding kinetics of p19 F86W monitored (a) and (b) by stopped-flow fluorescence, (c) and (d) by stopped-flow CD. (e) and (f) The respective folding kinetics of truncated p19 ANK3–5. (a), (c), and (e) Unfolding was initiated by a rapid change from 0 M to 6 M urea. Refolding was initiated by rapid dilution from (b) 6 M urea to 0.9 M urea, (d) 7 M urea to 1.1 M urea and (f) 5 M urea to 0.7 M urea. Data can be best described by (a)–(c) a double-exponential function or (d)–(f) a single-exponential function (red line). Deviations of the fits from the experimental data are given as residuals below each trace.

sensitivity and time-resolution.[8] They agree with the rates for the slow phase in Figure 3(c) (unfolding) and the monophasic decay in Figure 3(d) (refolding).

Fluorescence-detected refolding and unfolding were biphasic under all conditions between 0.5 M and 7.5 M urea (Figure 4(a)), and the apparent rate constants from the unfolding and refolding experiments in the transition region (2.2–3.5 M urea) coincided very well. This suggests that folding is reversible and an intermediate is populated during unfolding and refolding. The logarithms of the apparent rate constants of fast refolding and unfolding (log k_f^f and log k_u^f) vary linearly with the concentration of urea, and k_f^f and k_u^f extrapolate to about 200 s^{-1} at 0 M urea and to 1.9 s^{-1} at 8 M urea. The limbs of the second chevron in Figure 4(a), which corresponds to the urea-dependence of the lower apparent rate constants, extrapolate to

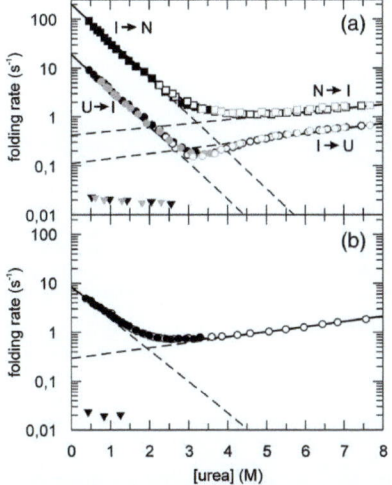

Figure 4. Urea-dependence of apparent folding rates of (a) p19 F86W and (b) p19 ANK3–5 monitored by fluorescence spectroscopy at 15 °C, pH 7.4. Filled symbols indicate refolding experiments, open symbols indicate unfolding experiments. (a) Folding rates of the fast phases from stopped-flow experiments are represented by squares and slow phases are represented by circles. Very slow refolding rates from manual mixing experiments are depicted as triangles. Gray symbols represent folding rates determined by stopped-flow CD. (b) All stopped-flow fluorescence unfolding and refolding experiments of p19 ANK3–5 were monophasic. A very slow phase was detected for p19 ANK3–5 by manual mixing experiments. The continuous line in (a) represents the result from a global analysis of kinetic and equilibrium data. Dotted lines represent the urea-dependence of the intrinsic rate constants for the indicated reaction. (b) The urea-dependence of the intrinsic unfolding and refolding rates are given as dotted lines according to a two-state model. The errors for the rate constants are smaller than the symbol.

$k_f^a = 19$ s^{-1} at 0 M urea and $k_u^a = 0.7$ s^{-1} at 8 M urea. Stopped-flow CD detected apparent rate constants (gray symbols in Figure 4(a)) correspond very well to the fluorescence data. k_f^f could not be resolved by this single-mixing method.

The unfolding and refolding kinetics were measured for the truncated variant without the two N-terminal repeats (p19 ANK3–5). It shows monophasic kinetics under all conditions (Figure 3(e) and (f)), and the measured rate constants followed a simple chevron with straight limbs at low and at high concentration of urea (Figure 4(b)). A linear extrapolation to 0 M urea gave values of $k_f = 8.2$ s^{-1} and $k_u = 0.29$ s^{-1}. The Tanford factor $\beta^T = m_f/(m_f - m_u)$ derived from the urea-dependences of log k_f and log k_u ($m_{UN}^{kin} = -3.53$ kJ/(mol·M) and $m_{NU}^{kin} = 0.59$ kJ/(mol·M)) is 0.86. This high value points to a native-like transition state.[25]

P19^{INK4d} shows in addition a very slow refolding reaction that is limited by prolyl isomerization.[8] This reaction was detected originally in real-time NMR experiments during refolding from 6 M urea to 2 M urea. It was observed also for p19^{INK4d} F86W and p19 ANK3–5 by both Trp fluorescence and CD (manual mixing). It shows a urea-independent rate constant of about 0.02 s^{-1} for both variants (triangles in Figure 4). This very slow refolding reaction is accelerated tenfold in the presence of equimolar prolyl isomerase SlyD(1–165) from *Escherichia coli*.[26] This confirms that the very slow refolding reaction is limited in rate by prolyl *cis/trans* isomerization.[27] It should be noted that all prolyl peptide bonds of the native state are in the *trans* conformation according to the crystal structure of p19^{INK4d}.[23,24]

Double-mixing experiments reveal a single kinetic folding intermediate

The biphasic kinetics suggest that unfolding as well as refolding of p19 F86W involve kinetic intermediates. To examine whether the same intermediate is formed, we correlated the unfolding and refolding kinetics by performing two sets of stopped-flow double-mixing experiments.

The first set of experiments (U assay) served to characterize the refolding kinetics of the intermediate that becomes populated during unfolding.[28,29] Native p19 F86W was first diluted from 0 M to 4.5 M urea to initiate unfolding. After short periods of time, unfolding was stopped and refolding started by a dilution to 1.5 M urea. The resulting refolding reactions were followed by Trp fluorescence (Figure 5). Under these conditions, the equilibrium-unfolded protein refolds in two phases, with rate constants of 13 s^{-1} and 1.8 s^{-1} (Figure 4(a)), and the fluorescence increases in both reactions. The same refolding kinetics were observed when, in the double-mixing experiment, unfolding was allowed to go to completion before switching to the refolding conditions in the second mixing step (Figure 5(a)).

A very different pattern was observed, however, after short unfolding times (Figure 5(a)). After 0.2 s of unfolding, refolding occurred in a single fast phase

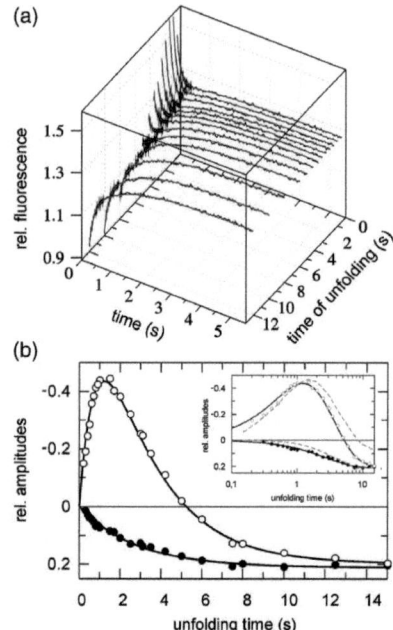

Figure 5. Double-mixing protein folding experiments of p19 F86W to monitor species during unfolding (U assay). Unfolding was initiated by a rapid change from 0 M to 4.5 M urea. After various times of unfolding, the subsequent refolding reaction under fluorescence detection was started by a second fast change to 1.5 M urea. (a) Double-exponential equations (red lines) were fit to the fluorescence intensities (continuous black lines). (b) The amplitudes from these fits are shown with open symbols for the fast folding phase and with closed symbols for the slow phase at different times of unfolding. A fit of a double-exponential function to the open symbols gave rate constants of $1.22(\pm 0.08)$ s^{-1} and $0.38(\pm 0.02)$ s^{-1}, the fit of a single-exponential function to the filled symbols gave $0.35(\pm 0.02)$ s^{-1} (continuous lines). The data were plotted on a logarithmic time-scale in the inset, where the dotted line represents a simulation of the U assay from the intrinsic unfolding and refolding rates derived from the global fit.

(13 s^{-1}), which was accompanied by a decrease in fluorescence (inset in Figure 5(b)). The magnitude of this fast decay increased rapidly with the time of unfolding and reached a maximal value after 1.3 s of unfolding (open symbols in Figure 5(b)). Then it decreased again, changed its sign after 5 s of unfolding and finally reached an equilibrium value of 0.2 after about 15 s of unfolding. With a small delay, a second, slow unfolding reaction appeared, which always showed a positive amplitude. This amplitude reached a limiting value of about 0.2 after 15 s (Figure 5(b)).

These double-mixing experiments reveal the relation between the intermediates detected in unfolding and refolding. The kinetic traces for the formation of the hyperfluorescent unfolding intermediate in Figure 3(a) and for the amplitude of fast refolding in Figure 5(b) are identical. This shows that the unfolding intermediate is in fact formed during the fast unfolding reaction (N→I, 1.2 s^{-1}), maximally populated after 1.3 s, and then converted into the unfolded form by the slow unfolding reaction (I→U, 0.4 s^{-1}). The data also demonstrate that the hyperfluorescent unfolding intermediate I refolds rapidly with a rate constant of 13 s^{-1}, which is identical with the rate of the fast refolding reaction at 1.5 M urea as measured in single-mixing experiments. The fast refolding reaction (the upper chevron in Figure 4(a)) thus reflects the formation of the native protein N from the intermediate form I. During unfolding, U is formed in a sequential reaction (N→I→U). This leads to a lag in the appearance of the slow refolding reaction (Figure 5(b)). The lag is small (about 0.2 s at 4.5 M urea) because the formation of I (N→I) is faster than its decay (I→U). This suggests also that the slow refolding reaction represents the U→I reaction (lower chevron in Figure 4(a)).

The second set of double-mixing experiments (N assay) provided complementary information on the refolding intermediate. In these experiments, unfolded protein was first diluted to 1.5 M urea to initiate refolding. Then, after various time intervals, refolding was interrupted by a second dilution to 6 M urea, and unfolding was monitored by fluorescence. Under these conditions, the native protein unfolds in two phases of opposite sign (cf. Figure 3(a)) with rate constants of 1.5 s^{-1} (N→I) and 0.6 s^{-1} (I→U). The two phases are also observed in the interrupted refolding experiments. The corresponding amplitudes depend on the duration of refolding in a complex manner (Figure 6(a)), which contains a wealth of information about the mechanism of refolding and the kinetic role of the refolding intermediate.

The filled symbols in Figure 6(a) represent the amplitude of the fast unfolding phase (N→I), and thus they trace the formation of the native state N during the refolding in the first step of the experiment. The open symbols show the amplitude of the slow unfolding phase (I→U) and thus trace the formation of both the I and the N molecules during refolding. The N molecules are also monitored, because N unfolds via I (N→I→U).

The dependence of the unfolding amplitudes on the duration of refolding in Figure 6(a) provides three important pieces of information on the folding mechanism of p19 F86W. Firstly, the fast phase with a rate constant of 15 s^{-1}, as observed during single-mixing refolding (cf. Figure 4(a)), is absent. This demonstrates that it is not a rate-limiting reaction for the formation of N during refolding. It rules out the possibility that the fast refolding reaction represents a parallel folding path to N and confirms that the two observed refolding phases represent sequential

Folding Mechanism of Ankyrin Repeat Protein p19INK4d

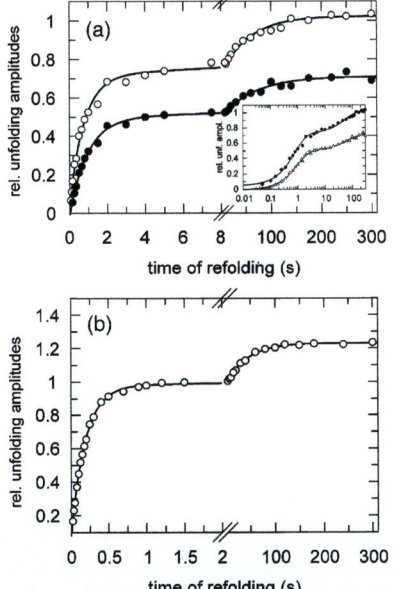

Figure 6. Double-mixing experiments to monitor intermediate and native species during refolding (N assay). (a) Amplitudes of the fast unfolding phase of p19 F86W at 6 M urea after various refolding times at 1.5 M urea are depicted by filled symbols and amplitudes of the slow unfolding phase are depicted by open symbols. A fit of a double-exponential function to the amplitude dependence upon refolding time (continuous lines) revealed for both data sets rate constants of $0.68(\pm 0.02)$ s^{-1} and $0.018(\pm 0.002)$ s^{-1}. The inset shows the same set of data on a logarithmic time-scale, where the dotted line represents the formation of the native state calculated by using the unfolding and refolding rates derived from the global fit. (b) Amplitudes of the mono-exponential unfolding of p19 ANK3-5 during the N assay at 0.45 M urea. The continuous line gave refolding rate constants of $5.2(\pm 0.07)$ s^{-1} and $0.022(\pm 0.001)$ s^{-1}.

reactions (Figure 4(a) at 1.5 M urea: U→I, 1.8 s^{-1} and I→N, 15 s^{-1}) on a single pathway. The U→I reaction is rate-limiting, and this explains why the formation of N molecules (Figure 6(a)) is governed by the rate constant of the U→I reaction.

The second piece of information arises for very short refolding times (≤ 0.1 s). Here, the unfolding kinetics are monophasic and show the slow (I→U) but not the fast unfolding reaction (N→I). The latter reaction appears with a lag of about 0.1 s, which is seen most clearly in the inset of Figure 6(a). The lag phase was observed in all N assay datasets, where the concentration of urea was varied during the refolding time (data not shown). This again confirms that in refolding N is formed in a sequential reaction via I, in which the formation of I is rate-limiting.

Thirdly, 77% of the N molecules are formed with the rate constant of the U→I reaction (1.8 s^{-1}). The remaining N molecules are formed slowly with a rate constant of 0.018 s^{-1} (Figure 6(a)). They originate from the unfolded molecules with non-native prolyl isomers. The amplitude of this reaction indicates that 23% of all unfolded molecules contain such proline residues. In good agreement with this, the above-mentioned real-time NMR experiment revealed 17% for the non-native prolyl isomers, a refolding rate constant of 0.017 s^{-1} and a 1D ^{1}H-NMR spectrum typical for an unfolded protein.[8] Addition of the prolyl isomerase *Escherichia coli* SlyD (1–165) to the refolding buffer during the N assay increased this very slow refolding phase (data not shown).

Similar double-mixing experiments were done with the variant lacking the first two ankyrin repeats (p19 ANK3-5). The single-mixing experiments (Figure 4(b)) suggested that the unfolding and refolding of this variant is a simple two-state reaction without an intermediate. The double-mixing experiments (Figure 6(b)) confirm this, and demonstrate that the native molecules are formed in two reactions: 80% of all molecules form the native state in a reaction with a rate constant of 5.2 s^{-1}, which is in good agreement with the rate constant of direct refolding; and 20% show non-native prolyl isomers and refold to the native state with a rate constant of 0.022 s^{-1}, also in good agreement with the previously measured value. It is likely that the rate-limiting proline residues are located in ankyrin repeats 3–5, because the full-length protein and the truncated form without the two N-terminal repeats show essentially the same fraction of molecules with non-native prolyl isomers. The rate constants of conversion and the populations are also similar to those measured for the Notch ankyrin domain, which contains six repeats and for which the role of individual proline residues has been studied in detail.[15]

The folding intermediate at equilibrium

The double-mixing experiments showed that the same kinetic intermediate I occurs in both unfolding and refolding. Here, we examined whether this intermediate is also populated at equilibrium. Our previous NMR experiments with p19^{INK4d} suggested that an intermediate is populated at equilibrium in the urea-induced transition.[8] A sensitive kinetic assay (I assay) was developed to relate the kinetic intermediate with the equilibrium intermediate. This assay makes use of the findings that the kinetic intermediate is strongly populated as a transient species during unfolding at high concentrations of urea, that it is hyperfluorescent, and that its unfolding can be followed very well by the decrease of its fluorescence (Figure 3(a)).

To search for I molecules by this assay, we equilibrated samples of p19 F86W at concentrations

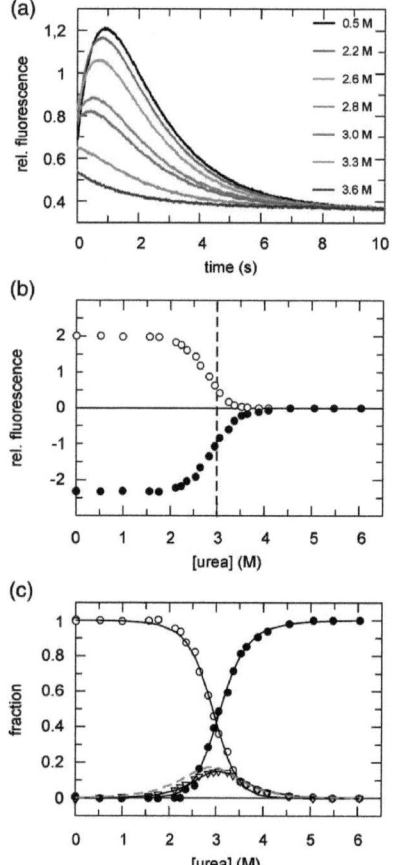

Figure 7. Assay for intermediates. Unfolding of p19 F86W was monitored at 6.6 M urea after equilibration of the protein at concentrations of urea between 0 M and 6 M. (a) Unfolding kinetics of p19 F86W incubated between 0.5 M and 3.6 M urea. (b) Amplitudes of the fast phase (1.48(±0.03) s^{-1}; open symbols) and the slow phase (0.53(±0.02) s^{-1}, filled symbols) of unfolding as a function of the concentration of urea used for equilibration. Double-exponential functions were fit to the measured kinetics. The dotted line represents 3 M urea close to the maximum population of I. (c) Calculated equilibrium populations of the native (open symbols), intermediate (triangles), and unfolded state (filled symbols). The dotted line represents the expected population of I from the urea-dependence of the intrinsic unfolding and refolding rate constants determined from a global fit and time extrapolated to equilibrium.

of urea ranging from 0 M to 6 M. These samples were unfolded under the same final conditions of 6.6 M urea and the time-courses of unfolding were followed by fluorescence. Representative kinetics are shown in Figure 7(a). The kinetics were biphasic with identical rate constants of 1.49(±0.03) s^{-1} for the N→I and 0.53(±0.02) s^{-1} for the I→U reaction. For the native protein, the amplitude of N→I is +2.0, and for the subsequent I→U reaction it is −2.3, as shown for unfolding of p19 F86W equilibrated at 0.5 M urea in Figure 7(a). The amplitudes of these two reactions decrease when the protein enters the transition region (Figure 7(b)). If only N and U molecules were present at equilibrium, the ratio of these two amplitudes would remain constant. This is clearly not the case. In the unfolding kinetics of the samples that were incubated in the transition region (2.6–3.6 M urea, Figure 7(a) and (b)) the amplitude of the fast unfolding reaction (N→I, open symbols in Figure 7(b)) gets lost at lower concentrations of urea than the amplitude of the slow unfolding reaction (I→U, filled symbols in Figure 7(b)). This is compelling evidence that I molecules exist at equilibrium in the transition region and contribute to the second (I→U), but not to the first (N→I) phase in the unfolding assay. The I molecules populated at equilibrium unfold with the same rate as I molecules formed transiently during the unfolding kinetics of the native protein. The asynchrony between the two unfolding amplitudes is seen most clearly by the missing over-shoot at 3.6 M urea (blue curve in Figure 7(a)) and between 3.0 M and 3.5 M urea in Figure 7(b).

The amplitudes from the kinetic unfolding assays (Figure 7(b)) allow an estimate of the fractions of N and U molecules in the transition region. The amplitude of the fast reaction (N→I) measures the concentration of N molecules. Its urea-dependence is shown in Figure 7(c) as open symbols (after normalization). The amplitude of the slow unfolding reaction (I→U) measures the concentration of all species that are not in the U state, and therefore the difference from the maximal amplitude (as observed between 0 M and 2 M urea, Figure 7(b)) reflects the concentration of the U molecules. The normalized curve for U is also shown in Figure 7(c) (filled symbols). In the transition region, N and U do not sum to 1. The difference (plotted in Figure 7(c) as triangles) reflects the population of the intermediate I under equilibrium, which shows a maximum of 15% at ∼3 M urea.

Global analysis of kinetic and equilibrium data

The analysis of the double-mixing experiments and the assay to detect the equilibrium intermediate substantiate Scheme 1 for the folding mechanism of p19^{INK4d}:

$$U \underset{k_{IU}}{\overset{k_{UI}}{\rightleftharpoons}} I \underset{k_{NI}}{\overset{k_{IN}}{\rightleftharpoons}} N$$

Scheme 1.

To further confirm that a single on-pathway intermediate can explain all kinetic and equilibrium observations, a global fit according to the mechanism of Scheme 1 has been done. The following kinetic and equilibrium data of p19 F86W were analyzed jointly: all unfolding and refolding kinetics detected by single-mixing stopped-flow fluorescence and CD experiments (Figure 4(a)), the CD-detected urea transition at equilibrium, and the final intensities of the fluorescence folding kinetics. The latter senses the increased fluorescence of the intermediate relative to U and N at concentrations of urea in the transition region, were I is populated significantly. The fitting protocol calculates the apparent folding rates from the intrinsic unfolding and refolding rate constants k_{NI} (N→I), k_{IU} (I→U), k_{IN} (I→N) and k_{UI} (U→I) at each concentration of urea according to a sequential three-state model for comparison with the observed kinetics.[25,30] From the four rate constants $k_{i,j}$, the Gibbs free energies between N and I (ΔG_{NI}) and between I and U (ΔG_{IU}) according to a three-state model[31] were calculated for comparison with the final fluorescence values and the CD urea transition. A linear urea-dependence of $\ln(k_{i,j}([urea])) = \ln(k_{i,j}^0) + m_{i,j}^{kin}$ [urea] for all four rate constants and the Gibbs free energies:

$$\Delta G_{i,j}([urea]) = \Delta G_{i,j}^0 + RT(m_{i,j}^{kin} - m_{j,i}^{kin})[urea]$$

was assumed. The fitting protocol minimized the deviation of all included apparent rate constants and equilibrium data from the simulated data according to Scheme 1 by globally adjusting $k_{i,j}^0$, $m_{i,j}^{kin}$, and the baselines of the equilibrium data. The results are given in Tables 1 and 2. The apparent rate constants according to the global fit are depicted in Figure 4(a) as a continuous line with very good agreement with the experimental data. The urea-dependence of the four $\bar{k}_{i,j}$ rates are plotted as dotted lines. The thermodynamic parameters derived from the global fit are listed in the row Kinetics of F86W in Table 1. ΔG_{NU}^0 and m_{NU} from the kinetic analysis agree well with the values from the equilibrium unfolding transitions.

The intrinsic rates derived from the global fit were used to calculate the amplitude profiles in the double-mixing experiments and the I assay. For the U assay (dotted gray line in the inset of Figure 5(b)), the over-shoot of the fast refolding amplitudes as well as the lag for the population of U could be reproduced. For the N assay (dotted gray line in the inset of Figure 6(a)), the calculated curve follows the experimental data very well, including the lag in the formation of N. The calculated population of the intermediate at equilibrium (dotted gray line in Figure 7(c)) also agreed well with the I assay. The gain in thermodynamic stability of p19^{INK4d} when progressing from the unfolded to the intermediate state (ΔG_{IU}^0 in Table 1) is comparable to ΔG_{NU}^0 of the truncated variant p19 ANK3–5. The m-values of this variant ($m_{UN}^{kin} = -3.53$ kJ/(mol·M) and $m_{NU}^{kin} = 0.59$ kJ/(mol·M)) also correspond well to the respective values m_{UI}^{kin} and m_{IU}^{kin} of full-length protein (Table 2). This indicates that the repeats ANK3–5 are already formed in the intermediate (for further evidence, see below).

Hydrogen/deuterium (H/D) exchange monitored by NMR spectroscopy

The kinetic analysis identified an on-pathway folding intermediate and gave values for the rates of its interconversion with U and N, and its stability against urea. It did not, however, give information about its structure. The comparisons between the full-length protein and the truncated variant without repeats 1 and 2 (Figure 4 and Table 1) suggested that in the intermediate repeats 1 and 2 are still unfolded or incompletely folded. Insight into the structural properties of folding intermediates can be gained, in principle, from mutational analysis. However, there is evidence that the folding mechanisms of ankyrin repeat proteins can change upon mutation,[11–13] which limits a validation of the wild-type mechanism if, for example, intermediates populate only in variants but not the wild-type protein.[10,11]

Therefore, we employed H/D exchange[32] as a sensitive tool to measure the local stability of wild-type p19^{INK4d} protein. Its amide resonances were assigned earlier.[33] The experimental exchange rate constants k_{ex} were determined on a residue-by-residue basis by following the single-exponential decay of cross-peak intensities in 48 ^1H-^{15}N heteronuclear single quantum coherence (HSQC) spectra acquired sequentially during 27 h after dissolving uniformly ^{15}N-labeled wild-type p19^{INK4d} in ^2H_2O at 15 °C. The protection factors $P = k_{int}/k_{ex}$ were derived by dividing the intrinsic exchange rate constants k_{int} determined from peptide models by k_{ex}.[34] They reveal marked differences in the stabilities of the individual repeats (Figure 8 and color coding in Figure 1). Repeats 3 and 4 show the highest stability and several amides in these repeats exchange only after global unfolding of the entire protein. P of these most protected residues around 2×10^5 correspond to ΔG_{NU}^0 of p19^{INK4d} (Table 1). The amides in repeat 5 are, on average, tenfold, and those in repeats 1 and 2 are ~100-fold less protected than those in repeats 3 and 4, indicating that the two N-terminal repeats of p19^{INK4d} undergo frequent local opening. These two repeats are missing in the p19 ANK3–5 variant, which does not show the I to N folding reaction (Figure 4(b)). Together, these results suggested that the two N-terminal repeats might

Table 2. Kinetic data derived from the global fit of the equilibrium transition and the chevron plot of p19 F86W

Reaction	$k_{i,j}^0$ [a] (s^{-1})	$m_{i,j}^{kin}$ [b] (kJ mol^{-1} M^{-1})
U→I	19.1	−4.09
I→U	0.11	0.57
I→N	203.6	−4.17
N→I	0.42	0.43

[a] Rate constants at 0 M urea.
[b] The given values are from the global fit multiplied by RT.

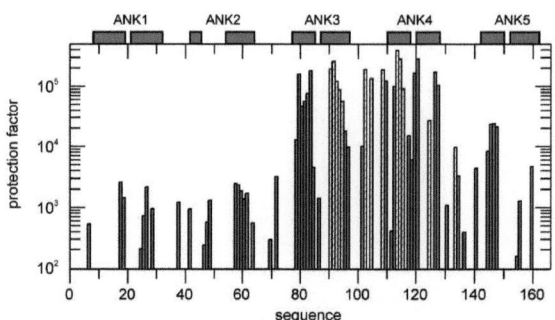

Figure 8. Protection factors of backbone amides of wild-type p19^{INK4d} determined by H/D exchange under NMR detection. Filled bars represent backbone amides, which were well protected, as determined during the first 90 min of H/D exchange. Missing bars indicate fast-exchanging amides, amides with missing assignments and proline. Hatched bars depict highly protected amides, which had exchanged by less than 10% after 27 h. Here, we assumed an experimental exchange rate below 0.095 day^{-1}, i.e. the given protection factor represents a lower limit.

still be unfolded or incompletely folded in the kinetic folding intermediate.

Discussion

Proteins composed of ankyrin repeats are widespread in nature. Most of them function as specific binding modules in cellular protein–protein recognition.[1] This property is being employed in biotechnology to generate ankyrin repeat proteins by *in vitro* evolution with high affinities for target proteins.[20–22] The linear arrangement of the repeats in these proteins suggests that they might unfold in a stepwise fashion. However, most of them show cooperative two-state equilibrium unfolding transitions.[14,16,18,19] On the other hand, complex folding kinetics were found for most ankyrin repeat proteins, including p16^{INK4a},[18] p19^{INK4d},[8] Notch ankyrin domain[9] and myotrophin.[11] Here, monophasic folding kinetics are the exceptions.[16]

Here, we showed that a linear three-state mechanism *via* an obligatory intermediate (Scheme 1) is necessary and sufficient to explain the unfolding and refolding kinetics and the equilibrium unfolding transition of p19^{INK4d}. A rigorous kinetic analysis of the folding mechanism of this protein was possible, because the fluorescent residue (W86) that was introduced artificially by the F86W mutation turned out to be an excellent probe for the formation of the folding intermediate without changing the stability and the folding mechanism. The fluorescence of W86 is quenched in the native protein, increases by more than fourfold during the N→I reaction and then decreases by sevenfold in the I→U reaction (at 6 M urea, Figure 3(a)). During refolding, the fluorescence of I is about fourfold above the final native state. Therefore, a double-exponential increase can be observed even in single-mixing experiments (Figure 3(c)), although the I→N reaction is tenfold faster compared to U→I.

The two folding reactions of p19^{INK4d} are separated from each other under all conditions, and they differ in rate by a factor of 10 in refolding and a factor of about 3 in unfolding. Usually, it is difficult to discriminate between parallel and sequential steps in complex folding kinetics. By using sets of double-mixing experiments we showed unambiguously that both unfolding (N→I→U) and refolding (U→I→N) are strictly sequential processes. This was possible because the rates of the two consecutive reactions are similar enough to produce significant lag phases in the formation of the final species. Additionally, the peculiar fluorescence properties of N, I and U provided very sensitive probes to follow the time-courses of the individual species in the double-mixing experiments. These experiments showed that the same intermediate is populated kinetically during unfolding and refolding as well as at equilibrium in the folding transition region.

The proposed folding mechanism in Scheme 1 was suggested for the Notch ankyrin domain.[9] Here, a lag in the minor unfolding phase confirmed that I is an on-pathway intermediate. For the four ankyrin repeats containing myotrophin, a sequential high-energy intermediate was proposed from a kink in the unfolding limb of the chevron plot of mono-phasic folding rates.[11] At low temperatures, a designed ANK protein comprising three ankyrin repeats followed a two-state model at equilibrium and in kinetic experiments.[16] This protein corresponds therefore to the p19 ANK3–5 fragment studied here.

When partially folded intermediates participate in the folding of small single-domain proteins, they usually form much faster than the rate-limiting step of folding, and therefore the two reactions are uncoupled. Such intermediates are typically much less stable than the native form, their kinetics cannot be resolved, and their existence at low concentrations of denaturant can be inferred only indirectly from a rollover in the chevron plot.[35] In rare cases such as Im7,[36] or the Notch ankyrin domain,[9] parts of the chevron plot representing the fast phases of an on-pathway intermediate could be resolved. For lysozyme, even more complex models considering four and five states of folding could be derived from multiphasic chevron plots.[37,38] In the case of p19^{INK4d} studied here, a complete biphasic chevron plot could be analyzed, because the formation and the decay of the intermediate (the U ↔ I reaction) are much slower than the interconversion with the native protein (I ↔ N). Moreover, the intermediate is kinetically always present, from 0 to 8 M urea.

A key to understanding this unusual folding mechanism and to structurally characterize the intermediate is provided by the folding kinetics of the truncated variant without the two N-terminal repeats and by the variations in the amide protection factors of the individual repeats. The truncated variant shows a stability and folding kinetics that resemble the U ↔ I transition of the intact protein, and, in the full-length protein, the repeats 3–5 show much higher amide protection factors than the repeats 1 and 2. Together, this suggests that the U → I reaction represents the folding of repeats 3–5 and that the I → N transition reflects the formation of the two N-terminal repeats and their docking with repeats 3–5. A φ-value analysis of p16^{INK4a} revealed a similar result, where ANK3 and ANK4 showed φ-values close to 1, which suggests that they are folded in the rate-limiting transition state of unfolding and refolding.[39]

The formation of the three C-terminal ANK repeats and the two N-terminal repeats are coupled events. Compared to the truncated variant (Figure 4(b)), the chevron plot representing the U ↔ I transition of full-length protein is shifted to the right by about 1 M urea, the unfolding limb is lowered by a factor of 2.5, and the refolding limb at low concentrations of urea is increased by a factor of 2. Since refolding is sequential, refolding as well as unfolding of the C-terminal repeats occur in the presence of the unfolded N-terminal repeats. For the N-terminal repeats the situation is different. They unfold and refold in the presence of the C-terminal repeats in the folded form. The latter are thus used as a scaffold for folding and for stabilizing the folded form of the N-terminal repeats. This folding on a scaffold explains why the N-terminal repeats can fold at all, and why they fold more rapidly than the C-terminal domain, despite their low level of intrinsic stability. The stability of the N-terminal repeats originates probably to a large extent from the stabilizing interactions with the C-terminal repeats. The low protection factors for the N-terminal repeats in the native state show that they can unfold while the C-terminal repeats remain folded. This corresponds to the first step of unfolding after changing the conditions to high concentrations of urea.

The fact that the kinetic intermediate is present also at equilibrium provides, in principle, a means for determining its structure. In fact, our previous analysis of the unfolding transition of p19^{INK4d} by NMR spectroscopy showed that I is populated to about 30% within the urea transition region, and several cross-peaks of I could be identified, but not assigned, primarily because the population of I is too low and the generally low dispersion of amide resonances in α-helical proteins.

The graded stability and the sequential folding mechanism are probably important for the function of p19^{INK4d}. Repeats 3–5 do not interact with CDK6, but form a stable assembly during refolding, which is then used by the repeats 1 and 2 as a scaffold for their own rapid refolding. This is accompanied by a further gain in Gibbs free energy and formation of the CDK6 binding interface, which resides on the repeats 1 and 2. Recently, it was found that the inhibitory role of p19^{INK4d} in the human cell-cycle is controlled by proteolysis. In vivo, μ-calpain cleaved selectively ankyrin repeats 1 and 2 of p19^{INK4d} and thus abolished inhibition of CDK4/6.[40] Calpains are Ca^{2+}-dependent, non-lysosomal, cysteine proteases that degrade several regulators of the G$_1$ to S transition.[41] We suggest that the low thermodynamic stability of repeats 1 and 2 and their facile unfolding are essential prerequisites for the cellular inactivation of p19^{INK4d} by regulatory proteolysis.

Materials and Methods

Protein expression and purification

P19^{INK4d} and mutants were prepared essentially as described but with minor modifications.[8,33] Mutations were introduced by using the Stratagene QuikChange Kit (Stratagene, La Jolla, CA). Deletion mutants were cloned in a pet15b vector, expressed and purified according to the full-length protein. Human CDK6 was expressed in SF9 insect cells infected with a glutathione-S-transferase (GST)-CDK6 baculovirus and purified as described.[23] GST pull-down assays were performed using standard protocols: 20 μl of GST-beads were incubated with purified GST-CDK6, washed, and added to 0.5 mg/ml solutions of p19^{INK4d} variants. After incubation for 1 h at 4 °C, beads were washed three times and then analyzed by SDS-PAGE.

Equilibrium CD and fluorescence spectroscopy

Urea was purchased from Gerbu and all other chemicals were from Merck. All experiments were performed at 15 °C and in 20 mM sodium phosphate (pH 7.4). CD spectra were recorded with a JASCO J600A spectropolarimeter (0.1 cm cell length, 10 μM protein concentration, 1 mm bandwidth) and corrected for the buffer contributions. Far-UV CD urea-induced unfolding transitions of p19^{INK4d} and mutants were monitored at 222 nm for 1–3 μM protein solutions with various concentrations of urea and incubation for 4–6 h to reach equilibrium. Urea transitions monitored by fluorescence were recorded with a Hitachi F-4010 and JASCO FP6500 fluorescence spectrometer with an excitation wavelength of 280 nm, an emission wavelength of 325–375 nm at a protein concentration of 1–3 μM. The experimental data were analyzed according to a two-state or three-state model[31] by non-linear, least-squares fit with proportional weighting to obtain the Gibbs free energy of denaturation ΔG as a function of the concentration of urea.

Kinetic single and double-mixing experiments

Fast kinetic experiments were performed using an Applied Photophysics SX-17MV and SX-20MV stopped-flow instrument at 15 °C. An excitation wavelength of

280 nm was used and emission was monitored at wavelengths above 305 nm using cut-off filters. Unfolding experiments were performed by mixing protein in 20 mM sodium phosphate (pH 7.4) with 10 volumes of urea in the same buffer. Refolding was initiated by 11-fold dilution of unfolded protein (5–7 M urea). For all p19^{INK4d} variants, the final concentration of protein was 1–3 µM. Data collected from four to eight scans were averaged and fit using GraFit 5 (Erithacus Software). Unfolding traces were analyzed by double-exponential functions, and refolding traces were analyzed by triple-exponential functions. The slowest refolding phase was determined by manual mixing and set constant for analysis of the fast refolding kinetics. For the I assay, 22 µM protein was incubated in various concentrations of urea (0–6 M) and unfolded into 6.6 M urea by 11-fold dilution. The resulting unfolding curves were analyzed according to double-exponential functions. Stopped-flow double-mixing experiments were performed for the N assay and for the U assay. For the measurement of unfolding kinetics (N assay), the unfolded protein (82.5 µM in 6 M urea) was first diluted 11-fold with buffer to allow refolding for 30 ms–300 s at 0.7 M–2.5 M urea, and subsequently diluted fivefold with urea solutions to obtain 1.5 µM protein in final concentrations of 4.5 M–6 M urea. For the U assay, native protein (82.5 µM) was first unfolded for various times (30 ms–30 s) at 4.5 M urea and subsequently refolded at 0.75 M–2.5 M urea by sixfold dilution.

Stopped-flow CD measurements were performed at 15 °C on an Applied Photophysics π*-180 instrument. Ellipticity was monitored at 225 nm in a 1 cm path-length cell. Unfolding and refolding experiments were performed as described for the stopped-flow fluorescence. After mixing, the protein concentration was 5 µM. Data from four to eight scans were averaged and analyzed. The apparent rate constants for unfolding and refolding were derived by fitting a double-exponential function. For the global fit with the program Scientist (MicroMath), as described in the main text, the analytical solutions of a three-state model were used.^{25,30} The fluorescence detected chevron plot was measured three times and all apparent rates were jointly fitted with all other equilibrium and kinetic data.

NMR measurements

NMR spectra were recorded at 15 °C on a Bruker Avance 600. The H/D exchange reaction was started by dissolving lyophilized, ^{15}N-labeled p19^{INK4d} in ^{2}H$_2$O buffer, containing 20 mM sodium phosphate (pD 6.9, pH meter reading), 25 mM NaCl, 25 mM KCl). A series of 48 ^{15}N HSQC spectra was recorded during an exchange time of 27 h. To obtain more information about the fast exchanging residues, a second H/D exchange reaction was performed recording 22 ^{15}N HSQC spectra within 3 h. Protection factors (P) were derived from $P = k_{int}/k_{ex}$ (EX2 regime), where k_{int} is the intrinsic exchange rate constant from peptide models,^{34} and k_{ex} is the observed rate constant obtained by fitting a single-exponential function to the intensity decay of amide cross-peaks in the series of ^{15}N HSQC spectra. The Gibbs free energy of complete unfolding for p19^{INK4d} of 27.8 kJ/mol (see Table 1) converts into a protection factor of 1×10^5 by $\Delta G_U = -RT \ln(1/P)$, which is slightly below the most protected amide protons observed. This discrepancy can be explained by the cis/trans isomerization of the unfolded state, which was observed also for other proteins,^{42,43} and the stabilizing effect of ^{2}H$_2$O.^{44}

Acknowledgements

We thank P. Rösch for NMR spectrometer time at 600 MHz and T. Kiefhaber for stopped-flow CD spectrometer time. This research was supported by a grant from the Deutsche Forschungsgemeinschaft (Ba 1821/3-1 and GRK 1026), the excellence initiative of the state Sachsen-Anhalt and Boehringer Ingelheim Fonds.

Supplementary Data

Supplementary data associated with this article can be found, in the online version, at doi:10.1016/j.jmb.2007.07.063

References

1. Mosavi, L. K., Cammett, T. J., Desrosiers, D. C. & Peng, Z. Y. (2004). The ankyrin repeat as molecular architecture for protein recognition. Protein Sci. 13, 1435–1448.
2. Morgan, D. O. (1995). Principles of CDK regulation. Nature, 374, 131–134.
3. Sherr, C. J. (1996). Cancer cell cycles. Science, 274, 1672–1677.
4. Thullberg, M., Bartkova, J., Khan, S., Hansen, K., Ronnstrand, L., Lukas, J. et al. (2000). Distinct versus redundant properties among members of the INK4 family of cyclin-dependent kinase inhibitors. FEBS Letters, 470, 161–166.
5. Thullberg, M., Bartek, J. & Lukas, J. (2000). Ubiquitin/proteasome-mediated degradation of p19^{INK4d} determines its periodic expression during the cell cycle. Oncogene, 19, 2870–2876.
6. Scassa, M. E., Marazita, M. C., Ceruti, J. M., Carcagno, A. L., Sirkin, P. F., Gonzzalez-Cid, M. et al. (2007). Cell cycle inhibitor, p19INK4d, promotes cell survival and decreasese chromosomal aberrations after genotoxic insult due to enhanced DNA repair. DNA Repair, 6, 626–638.
7. Ceruti, J. M., Scassa, M. E., Flo, J. M., Varone, C. L. & Canepa, E. T. (2005). Induction of p19INK4d in response to ultraviolet light improves DNA repair and confers resistance to apoptosis in neuroblastoma cells. Oncogene, 24, 4065–4080.
8. Zeeb, M., Rösner, H., Zeslawski, W., Canet, D., Holak, T. A. & Balbach, J. (2002). Protein folding and stability of human CDK inhibitor p19^{INK4d}. J. Mol. Biol. 315, 447–457.
9. Mello, C. C., Bradley, C. M., Tripp, K. W. & Barrick, D. (2005). Experimental characterization of the folding kinetics of the notch ankyrin domain. J. Mol. Biol. 352, 266–281.
10. Lowe, A. R. & Itzhaki, L. S. (2007). Biophysical characterisation of the small ankyrin repeat protein myotrophin. J. Mol. Biol. 365, 1245–1255.
11. Lowe, A. R. & Itzhaki, L. S. (2007). Rational redesign of the folding pathway of a modular protein. Proc. Natl. Acad. Sci. USA, 104, 2679–2684.
12. Street, T. O., Bradley, C. M. & Barrick, D. (2007). Predicting coupling limits from an experimentally determined energy landscape. Proc. Natl. Acad. Sci. USA, 104, 4907–4912.
13. Werbeck, N. D. & Itzhaki, L. S. (2007). Probing a moving target with a plastic unfolding intermediate of

an ankyrin-repeat protein. *Proc. Natl. Acad. Sci. USA,* **104**, 7863–7868.
14. Zweifel, M. E. & Barrick, D. (2001). Studies of the ankyrin repeats of the Drosophila melanogaster Notch receptor. 2. Solution stability and cooperativity of unfolding. *Biochemistry,* **40**, 14357–14367.
15. Bradley, C. M. & Barrick, D. (2005). Effect of multiple prolyl isomerization reactions on the stability and folding kinetics of the notch ankyrin domain: experiment and theory. *J. Mol. Biol.* **352**, 253–265.
16. Devi, V. S., Binz, H. K., Stumpp, M. T., Plückthun, A., Bosshard, H. R. & Jelesarov, I. (2004). Folding of a designed simple ankyrin repeat protein. *Protein Sci.* **13**, 2864–2870.
17. Plaxco, K. W., Simons, K. T. & Baker, D. (1998). Contact order, transition state placement and the refolding rates of single domain proteins. *J. Mol. Biol.* **277**, 985–994.
18. Tang, K. S., Guralnick, B. J., Wang, W. K., Fersht, A. R. & Itzhaki, L. S. (1999). Stability and folding of the tumour suppressor protein p16. *J. Mol. Biol.* **285**, 1869–1886.
19. Mosavi, L. K., Williams, S. & Peng, Z. Y. (2002). Equilibrium folding and stability of myotrophin: a model ankyrin repeat protein. *J. Mol. Biol.* **320**, 165–170.
20. Binz, H. K., Amstutz, P., Kohl, A., Stumpp, M. T., Briand, C., Forrer, P. et al. (2004). High-affinity binders selected from designed ankyrin repeat protein libraries. *Nature Biotechnol.* **22**, 575–582.
21. Schweizer, A., Roschitzki-Voser, H., Amstutz, P., Briand, C., Gulotti-Georgieva, M., Prenosil, E. et al. (2007). Inhibition of caspase-2 by a designed ankyrin repeat protein: specificity, structure, and inhibition mechanism. *Structure,* **15**, 625–636.
22. Zahnd, C., Wyler, E., Schwenk, J. M., Steiner, D., Lawrence, M. C., McKern, N. M. et al. (2007). A designed ankyrin repeat protein evolved to picomolar affinity to Her2. *J. Mol. Biol.* **369**, 1015–1028.
23. Brotherton, D. H., Dhanaraj, V., Wick, S., Brizuela, L., Domaille, P. J., Volyanik, E. et al. (1998). Crystal structure of the complex of the cyclin D-dependent kinase Cdk6 bound to the cell-cycle inhibitor p19^{INK4d}. *Nature,* **395**, 244–250.
24. Russo, A. A., Tong, L., Lee, J. O., Jeffrey, P. D. & Pavletich, N. P. (1998). Structural basis for inhibition of the cyclin-dependent kinase Cdk6 by the tumour suppressor p16^{INK4a}. *Nature,* **395**, 237–243.
25. Bachmann, A. & Kiefhaber, T. (2005). Kinetic mechanisms in protein folding. In *Protein Folding Handbook* (Buchner, J. & Kiefhaber, T., eds), vol. 1, pp. 379–406. Wiley-VCH, Weinheim.
26. Scholz, C., Eckert, B., Hagn, F., Schaarschmidt, P., Balbach, J. & Schmid, F. X. (2006). SlyD proteins from different species exhibit high prolyl isomerase and chaperone activities. *Biochemistry,* **45**, 20–33.
27. Balbach, J. & Schmid, F. X. (2000). Proline isomerization and its catalysis in protein folding. In *Mechanisms of Protein Folding* (Pain, R. H., ed.), 2nd edit., pp. 212–237, University Press, Oxford.
28. Nall, B. T., Garel, J.-R. & Baldwin, R. L. (1978). Test of the extended two-state model for the kinetic intermediates observed in the folding transition of ribonuclease A. *J. Mol. Biol.* **118**, 317–330.
29. Schmid, F. X. & Baldwin, R. L. (1978). Acid catalysis of the formation of the slow-folding species of RNase A: evidence that the reaction is proline isomerization. *Proc. Natl. Acad. Sci. USA,* **75**, 4764–4768.
30. Ikai, A. & Tanford, C. (1973). Kinetics of unfolding and refolding of proteins. I. Mathematical analysis. *J. Mol. Biol.* **73**, 145–163.
31. Hecky, J. & Müller, K. M. (2005). Structural perturbation and compensation by directed evolution at physiological temperature leads to thermostabilization of β-lactamase. *Biochemistry,* **44**, 12640–12654.
32. Englander, S. W., Downer, N. W. & Teitelbaum, H. (1972). Hydrogen exchange. *Annu. Rev. Biochem.* **41**, 903–924.
33. Kalus, W., Baumgartner, R., Renner, C., Noegel, A., Chan, F. K., Winoto, A. & Holak, T. A. (1997). NMR structural characterization of the CDK inhibitor p19^{INK4d}. *FEBS Letters,* **401**, 127–132.
34. Bai, Y. W., Milne, J. S., Mayne, L. & Englander, S. W. (1993). Primary structure effects on peptide group hydrogen exchange. *Proteins: Struct. Funct. Genet.* **17**, 75–86.
35. Capaldi, A. P., Kleanthous, C. & Radford, S. E. (2002). Im7 folding mechanism: misfolding on a path to the native state. *Nature Struct. Biol.* **9**, 209–216.
36. Capaldi, A. P., Shastry, M. C., Kleanthous, C., Roder, H. & Radford, S. E. (2001). Ultrarapid mixing experiments reveal that Im7 folds via an on-pathway intermediate. *Nature Struct. Biol.* **8**, 68–72.
37. Wildegger, G. & Kiefhaber, T. (1997). Three-state model for lysozyme folding: triangular folding mechanism with an energetically trapped intermediate. *J. Mol. Biol.* **270**, 294–304.
38. Bieri, O. & Kiefhaber, T. (2001). Origin of apparent fast and non-exponential kinetics of lysozyme folding measured in pulsed hydrogen exchange experiments. *J. Mol. Biol.* **310**, 919–935.
39. Tang, K. S., Fersht, A. R. & Itzhaki, L. S. (2003). Sequential unfolding of ankyrin repeats in tumor suppressor p16. *Structure,* **11**, 67–73.
40. Joy, J., Nalabothula, N., Ghosh, M., Popp, O., Jochum, M., Machleidt, W. et al. (2006). Identification of calpain cleavage sites in the G1 cyclin-dependent kinase inhibitor p19(INK4d). *Biol. Chem.* **387**, 329–335.
41. Goll, D. E., Thompson, V. F., Li, H., Wei, W. & Cong, J. (2003). The calpain system. *Physiol. Rev.* **83**, 731–801.
42. Bai, Y. W., Milne, J. S., Mayne, L. & Englander, S. W. (1994). Protein stability parameters measured by hydrogen exchange. *Proteins: Struct. Funct. Genet.* **20**, 4–14.
43. Steegborn, C., Schneider-Hassloff, H., Zeeb, M. & Balbach, J. (2000). Cooperativity of a protein folding reaction probed at multiple chain positions by real-time 2D NMR spectroscopy. *Biochemistry,* **39**, 7910–7919.
44. Sato, S. & Raleigh, D. P. (2007). Kinetic isotope effects reveal the presence of significant secondary structure in the transition state for the folding of the N-terminal domain of L9. *J. Mol. Biol.* **370**, 349–355.
45. Koradi, R., Billeter, M. & Wüthrich, K. (1996). MOLMOL: a program for display and analysis of macromolecular structures. *J. Mol. Graph.* **14**, 51–55.
46. Baumgartner, R., Fernandez-Catalan, C., Winoto, A., Huber, R., Engh, R. A. & Holak, T. A. (1998). Structure of human cyclin-dependent kinase inhibitor p19^{INK4d}: comparison to known ankyrin-repeat-containing structures and implications for the dysfunction of tumor suppressor p16^{INK4a}. *Structure,* **6**, 1279–1290.

Supplementary material

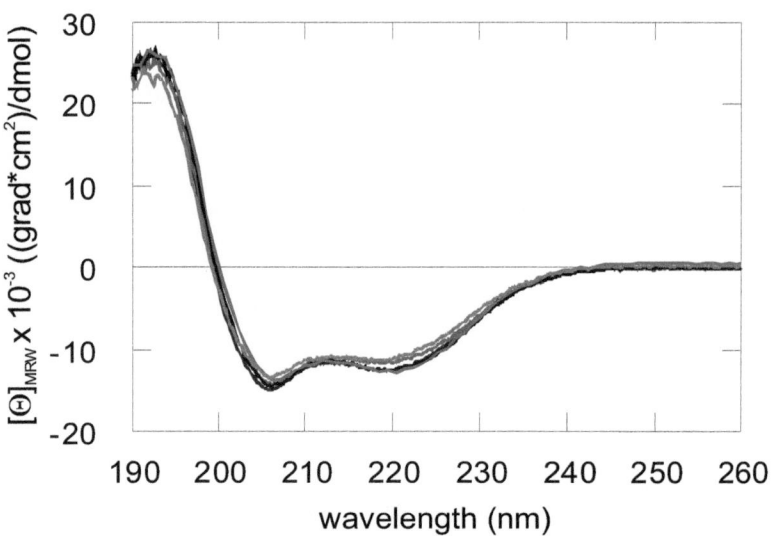

Figure S1 Far UV-CD spectra of p19^{INK4d} (black), p19 F86W (blue), p19 ANK3-5 (green), and p19 ANK1-4 (red) given as mean residue weighted ellipticity superimpose between 190 nm and 260 nm at 15°C, pH 7.4.

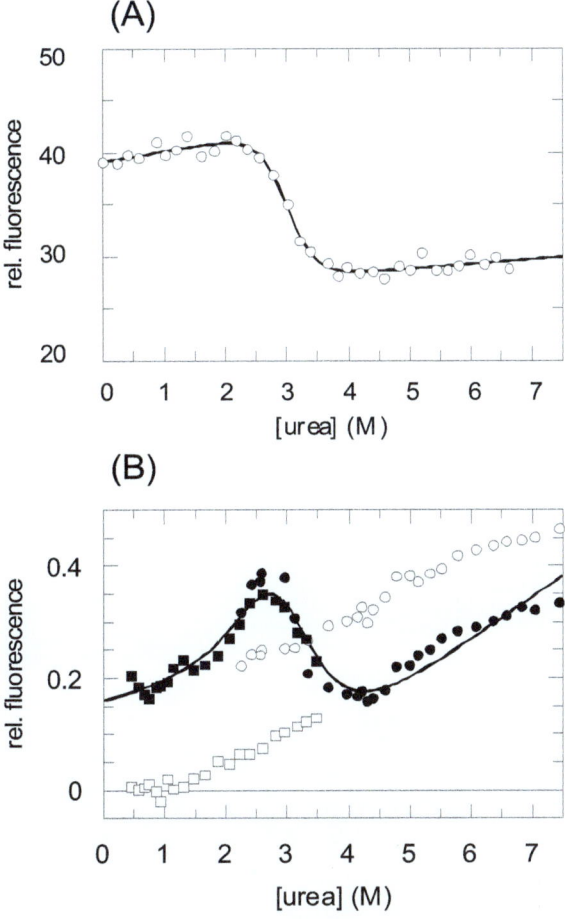

Figure S2 (A) Urea equilibrium transitions of p19 F86W monitored at 320 nm by fluorescence spectroscopy at 15°C. It can be best described by a two state model (solid line). (B) Initial and final value analysis of p19 F86W resulting from single mixing stopped flow fluorescence kinetics. Initial fluorescence intensities for unfolding (circles) and refolding (squares) reactions are illustrated by open symbols. Closed symbols represent the final fluorescence of unfolding (circles) and refolding kinetics (squares). The urea dependence of the final values follows a three state behavior (solid line) and was analyzed together with kinetic data (see global fit section in the main text).

7.2 Subproject B

B

Subproject B

Subproject B

Classification: Biological Sciences, Biophysics

Conformational Switch upon Phosphorylation: Human CDK Inhibitor p19^{INK4d} Between Native and Partially Folded State

Christian Löw[1], Nadine Homeyer[2], Ulrich Weininger[1], Heinrich Sticht[2], Jochen Balbach[*,1,3]

[1] Institut für Physik, Biophysik, Martin-Luther-Universität Halle-Wittenberg, D-06120 Halle (Saale), Germany

[2] Abteilung Bioinformatik, Institut für Biochemie, Friedrich-Alexander-Universität Erlangen-Nürnberg, 91054 Erlangen, Germany

[3] Mitteldeutsches Zentrum für Struktur und Dynamik der Proteine (MZP), Martin-Luther-Universität Halle-Wittenberg, Germany

*Correspondence should be addressed to

Jochen Balbach	Tel.:	++49 345 55 25353
Institut für Physik, Biophysik	FAX:	++49 345 55 27383
Martin-Luther-Universität Halle-Wittenberg	e-mail:	jochen.balbach@physik.uni-halle.de
D-06120 Halle(Saale), Germany		

Manuscript information: 19 text pages, 165 words (abstract), 39111 characters (text), 9325 characters (estimated for 7 figures)

Abbreviations: AR, ankyrin repeat; CDK, cyclin dependent kinase; MD simulation, molecular dynamic simulation; CKI, cyclin dependent kinase inhibitors; HSQC, heteronuclear single quantum coherence.

Abstract

$P19^{INK4d}$ (p19) consists of five ankyrin repeats (AR) and controls the human cell cycle by inhibiting the cyclin D-dependent kinases 4 and 6. Posttranslational phosphorylation of p19 has been described for S66 and S76. In the present study we show that mimicking the phosphorylation site of p19 by a glutamate substitution at position 76 dramatically decreases the stability of the native but not the intermediate state. At body temperature the native conformation is completely lost and p19 molecules exhibit the intermediate state as judged by kinetic and equilibrium analysis. High resolution NMR spectroscopy verified that the three C-terminal repeats remain folded in the intermediate state, whereas all cross-peaks of the two N-terminal repeats lost their native chemical shift. Molecular dynamic simulations of p19 in different phosphorylation states revealed large scale motions in phosphorylated p19, which cause destabilization of the interface between the second and third AR. Double phosphorylated p19 mimic mutants showed an increased accessibility for ubiquitination indicating a direct link between phosphorylation and ubiquitination.

Keywords: ankyrin repeat; NMR; phosphorylation mimic; MD simulation; ubiquitination; cell cycle

Introduction

Cyclins and cyclin-dependent kinases (CDKs) are key regulators of the cell cycle (1, 2). They phosphorylate the retinoblastoma tumour suppressor (pRB) thereby inhibiting its growth-suppressive function and triggering an E2F-dependent transcriptional program that is necessary for completion of G1 and entering the S-phase of the cell cycle (3). But initiation of pRB phosphorylation strongly depends on the accumulation of D-type cyclins and their assembly with CDK4 or CDK6 (4). The activity of these holoenzymes is further regulated by phosphorylation of CDKs, proteolysis of cyclins, and specific inhibitory proteins named cyclin dependent kinase inhibitors (CKIs) (5, 6). Two families of CKIs are known. CKIs of the INK4 family ($p16^{INK4a}$, $p15^{INK5b}$, $p18^{INK4c}$ and $p19^{INK4d}$) specifically bind and inhibit the corresponding kinases of the D-type cyclins, CDK4 and CDK6, whereas CKIs of the Cip/Kip family (p21, p27 and p57) inhibit a broader spectrum of CDKs (1, 2, 6-8). The four members of the INK4 family share a similar protein fold consisting either of four ($p15^{INK5b}$, $p16^{INK4a}$) or five ($p18^{INK4c}$, $p19^{INK4d}$) ankyrin repeats (AR). Characteristic for all members is helix two of the second AR, the latter consists of just one helical turn compared to the canonical AR fold (9-13).

Gene deletion, transcriptional silencing by promotor methylation, or mutations which inactivate CKIs are commonly found in diverse types of cancer and therefore attribute them to tumour suppressor proteins (14-17). Although the INK4 members appear structurally redundant and equally potent as inhibitors, a number of non-overlapping features have been described. They differ regarding their expression pattern during development and some of them participate in other fundamental processes such as DNA-repair, terminal differentiation, and cellular aging or senescence (18-21). The INK4 members $p15^{INK5b}$ (p15), $p16^{INK4a}$ (p16), and $p18^{INK4c}$ (p18) exhibit a remarkable low thermodynamic stability *in vitro* as shown by urea and GdmCl transitions (22-26) and protein half-lives in cell lines between four to six hours (18). In contrast, $p19^{INK4d}$ (p19) was found thermodynamically more than twice as stable as p16 *in vitro*, but is rapidly degraded *in vivo* with a protein half-life of 20 to 30 minutes (18). It was shown, that the periodic oscillation of p19 during the cell cycle is determined by the ubiquitin/proteasome dependent mechanism, which appears to be restricted to p19 within the INK4 family. Lysine 62 of p19 was found to be targeted by ubiquitination (18). Analysis of further posttranslational modifications including phosphorylation, which is known to regulate function, subcellular localization, and turnover of diverse cell cycle regulatory proteins including the CKIs of the Cip/Kip family, revealed a differential phosphorylation pattern for the INK4 proteins *in vivo*. No phosphorylation was seen for p15 and 16, while p18 showed a detectable and p19 an even stronger phosphorylation signal. Single and double phosphorylated p19 was found in U-2-OS cell lines and identified phosphorylation sites were assigned to serines 66 and 76 so far. Both residues are conserved in human and mouse p19 but not throughout the entire INK4 family. These findings suggested a novel mechanism for controlling at least some aspects of p19 function and differentiate p19 from p15 and p16 (19).

Post-translational phosphorylation is an ubiquitous mechanism for cellular regulation (27). It is a key step in cell cycle control, gene regulation, transport, and metabolism (28). Nearly one third of the proteins in mammalian cells are expected to be phosphorylated at a given time point and the number of identified phosphorylation sites is growing quickly (29, 30). Modulation of protein activity or protein-protein interaction by phosphorylation is rapid and reversible and therefore advantageous for the cell, because it does not require the production of new proteins or the degradation of existing proteins. Although a major part of research focuses on post-translational phosphorylation, the detailed role of specific phosphorylation sites in proteins is often poorly understood. High resolution information obtained by X-ray crystallography or nuclear magnetic resonance is limited, since high amounts of homogenous purified protein in a specific modification state are difficult to achieve. Artificial post translational modification mimics are widely used to overcome this problem. Phosphorylation sites are commonly mimicked by Glu or Asp mutations (31-33), because a negative charge at the respective position can often approximate the function of the modified protein.

Currently, AR proteins are frequently used as models for protein folding studies. The simple modular architecture of the proteins predicts high folding rates. Experimentally, however, most AR proteins were found to fold much slower because of the slow formation of protein folding intermediates. Folding of p19 is best described by a three-state model, which involves the formation of an on-pathway intermediate as rate limiting step (22). High resolution information of intermediate states of AR proteins are difficult to accomplish, because wild type proteins typically show two-state unfolding transitions at equilibrium although folding kinetics are complex (34). In the present study we show that mimicking the S76 phosphorylation site of p19 by a glutamate substitution dramatically decreases the stability of the native state but not the on-pathway intermediate. At body temperature the p19 variant looses its native conformation and converts completely into the intermediate state as judged by kinetic and equilibrium analysis. High resolution NMR spectroscopy verified that the C-terminal repeats remain folded in the intermediate state, whereas all nuclei of the N-terminal repeats lost their native chemical shift. MD simulations of p19 in different phosphorylation states support the experimental observation that the introduction of a negative charge at position 76 has a significant effect on protein stability. Furthermore, the double phosphorylation mimic mutant (p19 S76E/S66E) showed an increased accessibility for ubiquitination indicating a direct link between phosphorylation and ubiquitination.

Results and Discussion

Urea induced unfolding involves the formation of a hyperfluorescent intermediate.

CDK inhibitor p19 consists of five ARs. It blocks the cell cycle at the transition of the G1- to the S-phase by binding and thus inhibiting cyclin dependent kinases 4 and 6. According to the crystal structure, binding to CDK6 is mainly mediated by AR 1 and 2. For folding studies, a fluorescence sensitive probe was introduced by replacing Phe 86 with a tryptophan residue, since p19 is devoid of any fluorophores (pseudo wild type). Stability and function of this mutant remained as observed for the wild type (22). Therefore, all further p19 mutants contained this mutation. Characteristic for the pseudo wild type protein is a hyperfluorescent intermediate, detectable in unfolding and refolding kinetics, but not under equilibrium conditions.

P19 is phosphorylated *in vivo* but the specific kinase is still unknown. Therefore, we mimicked the earlier identified phosphorylation sites at position 66 and 76 by glutamate substitutions (Fig. 1) to study the role of phosphorylation on stability, kinetics, and function. All p19 mutants were expressed in soluble form in *E. coli* except mutants with a serine to glutamate substitution at position 76. These mutants accumulated as inclusion bodies, but could be refolded successfully *in vitro*.

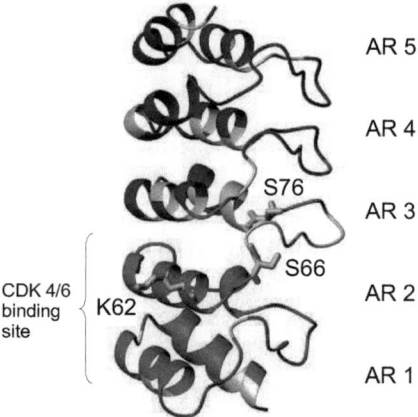

Fig. 1. Schematic representation of the crystal structure of p19^{INK4d} (1bd8.pdb from the Protein Data Base). Five ankyrin repeats (AR 1-5), each comprising a loop, a β-turn and two sequential α-helices form the elongated structure, where S66, S76, and K62 are indicated by a stick illustration of the side chains. Residues of the phosphorylation mimic mutant p19 S76E with native chemical shift at 37 °C are color coded in blue, while residues in red lost the native structure at body temperature. Indicated in grey are proline residues and residues which could not been evaluated due to signal overlap or missing assignment. The CDK4/6 binding site is mainly formed by the N-terminal AR 1 and 2. The figure was created using MOLMOL (52).

Urea induced unfolding transitions monitored by tryptophan fluorescence revealed that mutants containing glutamate at position 76 are strongly destabilized compared to the wild type protein. Furthermore, the glutamate phosphorylation mimic at position 66 or a serine to alanine substitution at position 76 reduces the stability only marginally (Table 1). Denaturation curves of all mutants monitored at 325 nm displayed two-state behaviour (Fig. 2A). A two-state analysis would artificially result in a strongly reduced *m*-value for the Glu 76 containing mutants compared to the pseudo wild type. Unfolding curves obtained by plotting the fluorescence intensity at higher wavelengths against the urea concentration, contained more information and show three-state behavior for the Glu 76 variant (Fig. 2B). These data could be fitted globally (at least eight transition curves between 320 - 380 nm) to a three-state model for the transition between the native state (N), intermediate (I), and the unfolded state (U). This hyperfluorescent intermediate state which was already seen in folding and unfolding kinetics is now significantly populated under equilibrium conditions (Fig. 2D).

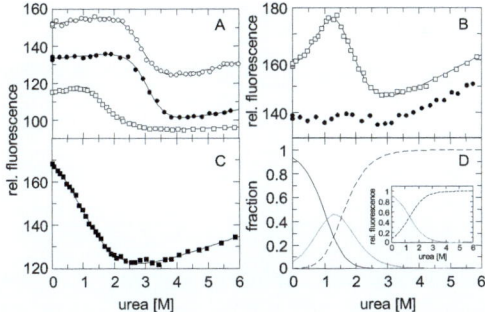

Fig. 2. Urea-induced unfolding of p19^{INK4d} mutants monitored by tryptophan fluorescence. Transition curves of p19 S66E (○), p19 S76E (□) and p19 S76A (●) at an emission wavelength of (A) 325 nm and (B) 375 nm at 15 °C. (C) Unfolding of p19 S76E/S66E (■) at 37 °C. Solid lines in (A-C) represent the least square fit of a two-state or three-state model. (D) Calculated equilibrium populations for the p19 S76E mutant of the native N (black line), intermediate I (grey line) and unfolded state U (dotted black line) according to the global analysis of the fluorescence equilibrium data at 15 °C. Inset shows the population profile for the same mutant at 37 °C.

While the native state is strongly destabilized, the stability of the intermediate state is less influenced (Table 1). Furthermore, global analysis resulted in similar m-values for all p19 mutants (within errors), suggesting that the mutations did not change the folding mechanism but the stability. By raising the temperature from 15 °C to body temperature (37 °C), equilibrium folding transitions of S76E containing p19 mutants simplified to a two-state mechanism (Fig. 2C). The fluorescence intensity was quenched upon addition of urea, indicating that the denaturation curve displays the transition from the hyperfluorescent intermediate state to the unfolded state. The resulting m-value was similar to the m-value obtained for the I to U transition at 15 °C rather than the N to U transition. Therefore, we conclude that p19 S76E resides in the I state at 37 °C.

Folding kinetics of p19 phosphorylation mimic mutants

A similar but more detailed picture was obtained by analyzing the unfolding and refolding kinetics of p19 S66E/S76E measured by stopped flow fluorescence spectroscopy. Unfolding and refolding of p19 is a biphasic process coupled to a slow prolyl *cis/trans* isomerization reaction in the unfolded state (omitted for clarity in this study). A detailed analysis of the folding mechanism was reported recently (22). Unfolding and refolding kinetics of the phosphorylation mimic mutants displayed the same biphasic properties as the pseudo wild type protein at 15 °C (Fig. 3A), characterized by the hyperfluorescent intermediate state. The fluorescence decay of the hyperfluorescent intermediate state in the unfolding reaction was assigned earlier to the I to U transition (22). The reaction from N to I is accompanied by a fluorescence increase. Refolding rates of p19 S66E/S76E for the I to N transition are significantly slower compared to wild type, while unfolding rates increased, resulting in the reduced stability for the N to I transition (Fig 3B). The native population is completely lost, when the temperature is raised to 37 °C as seen in equilibrium transitions. Folding kinetics displayed two-state behaviour, representing the I to U transition. Unfolding kinetics at 37 °C followed a single exponential decay without any detectable intermediates, clearly showing that the hyperfluorescent intermediate and not the native state is the starting point of the unfolding reaction. A similar folding behavior has been reported earlier for a deletion construct of p19, comprising AR 3-5 (22).

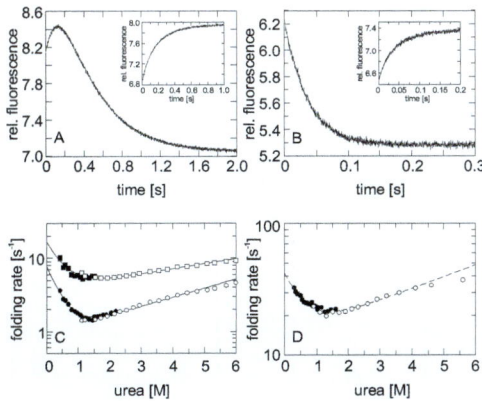

Fig. 3. Single mixing unfolding and refolding kinetics of p19 S76E/S66E detected by stopped flow fluorescence. Experimental data are plotted in black and fits in grey. Unfolding was initiated by a rapid change from 0 M to 3.2 M urea at 15 °C (A) and 37 °C (B) and can be best fitted by a double or single exponential function, respectively. Insets show fast refolding kinetics of the latter p19 mutant from 4.4 M to 0.4 M urea at the given temperature. The slowest refolding phase caused by prolyl *cis/trans* isomerisation is omitted for clarity. (C, D) Urea dependence of apparent folding rates of p19 S76E/S66E monitored at 15 °C and 37 °C. Closed symbols (●,■) represent folding rates of refolding experiments, open symbols (○,□) of unfolding experiments. Chevron plots were analyzed according to a three state (solid line) or two state model (dashed line).

Characterization of the intermediate state by NMR

As a result of the population of the intermediate state under equilibrium conditions caused by the phosphorylation mimic, it was possible to further structurally characterize this state by NMR spectroscopy. More than 82 % of the assignment of the backbone amide protons of the wild type protein could be directly transferred to the S76E mutant at 15 °C. Unfolding of native molecules towards the intermediate state was followed by a series of 15 2D ^{15}N-

TROSY-HSQC spectra recorded at temperatures between 15 °C and 40 °C. Cross-peaks of native AR 1-2 vanished at 37 °C whereas AR 3-5 remained folded with native-like chemical shifts (Fig. 1 and 4). ^{15}N- TROSY-HSQCs of the wild type protein were unchanged between 15 °C and 40 °C, which agrees well with an unfolding midpoint of 52 °C derived from CD unfolding transitions.

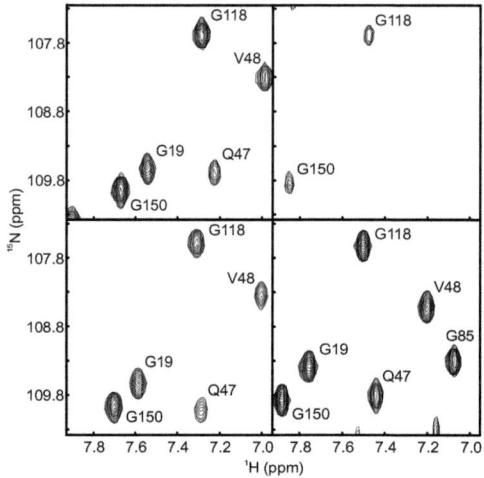

Fig. 4. Sections of ^{15}N-TROSY-HSQC spectra of p19 S76E (top panels) and p19 wild type (bottom panels) at 15 °C (left panels) and 37 °C (right panels). Native cross-peaks of AR 1-2 of p19 S76E vanished at 37 °C, whereas AR 3-5 still display native chemical shifts under these conditions. ^{15}N-TROSY-HSQC spectra of the wild type protein do not change between 15 °C and 37 °C. This shows that AR 3-5 of the phosphorylation mimic mutant p19 S76E remain folded in the intermediate state.

Thus, this high resolution NMR data of the p19 folding intermediate confirmed the earlier proposed scaffold function of AR 3-5 (blue in Fig. 1) for the less stable but functional AR 1-2 (red in Fig. 1). Phosphorylation mimic at position 76 strongly destabilized and uncoupled the native state from the intermediate state, leading to the population of the intermediate state under equilibrium conditions.

MD simulations of phosphorylated p19

As a complement to the preceding experimental results, molecular dynamics simulations were carried out with various phosphorylated p19 molecules to get detailed insights of structural and dynamic consequences, caused by the phosphorylations. MD simulations were performed for unphosphorylated, S66-phosphorylated (S66-P), S76-phosphorylated (S76-P) and S66-P/S76-P double phosphorylated p19.

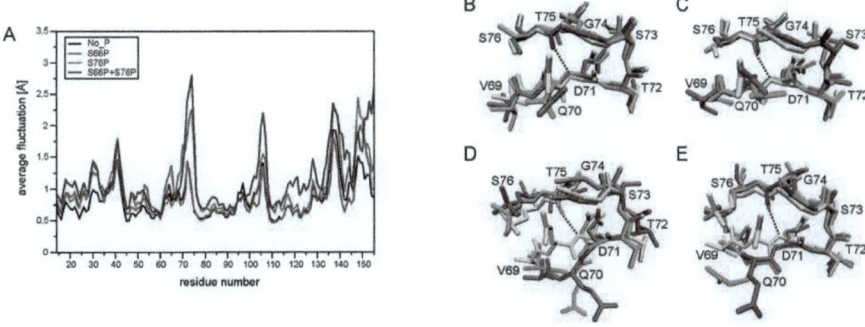

Fig. 5. Structural consequences of p19 phosphorylation on protein dynamics as deduced from 20 ns MD simulations. (A) Average fluctuations per residue, calculated over the backbone atoms of residues 14 - 155 for the 0 - 20 ns simulation period. The terminal residues which were either absent from the crystal starting structure or exhibited large fluctuations in the simulations were excluded from analysis. (B-E) Structural changes in the region of the S76 phosphorylation site. Overlay of starting structure (yellow), and structures recorded after 10 ns (red) and 20 ns (green) MD simulation of unphosphorylated (B), S66-P (C), S76-P (D), and S76-P/S66-P (E) p19.

In each simulation, the overall fold of p19 proved to be stable over the first 20 ns, and only local fluctuations were observed. The magnitude of these fluctuations, however, differed significantly between the simulations (Fig. 5A). While S66 phosphorylation leads only to a minor increase of the dynamics compared to unphosphorylated p19, the S76-P and S66-P/S76-P simulations showed significantly increased dynamics. This finding is in good agreement with our experimental data for the phosphorylation mimicking mutants, showing that a negative charge at position 76 has a large effect on the protein stability and that this effect is significantly larger compared to the introduction of a negative charge at position 66.

A detailed analysis of the simulations revealed that phosphorylation of S76 affects the hydrogen bonding pattern of the adjacent residues. In particular, a backbone hydrogen bond between the carbonyl oxygen of T75 and the amide group of D71 is lost in the S76-P and S66-P/S76-P simulations (Fig. 5B-E), thus explaining the higher flexibility of this loop in the respective simulations.

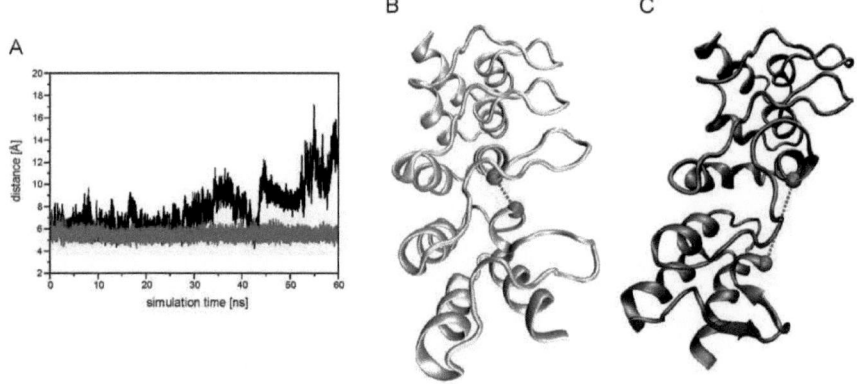

Fig. 6. Larger scale motions in double-phosphorylated p19 detected in longer MD simulations. (A) Q47(Cα) – P77(Cα) distance in the course of the 60 ns MD simulation of the double phosphorylated (S66-P/S76-P, black) and unphosphorylated (red) form of p19. (B, C) Comparison of the initial structure (B) and the structure of S66-P/S76-P p19 after 60 ns simulation time (C). The positions and the distance of Q47(Cα) and P77(Cα) are marked to highlight the structural changes in the interface between the second and third AR.

In order to investigate the effect of phosphorylation on longer time scales, we extended the simulations of unphosphorylated and double phosphorylated p19 for additional 40 ns leading to 60 ns overall simulation time. Consistent with the experimental data, the unphosphorylated protein remained stable over the entire simulation, while larger scale motions started in the double phosphorylated protein after 30 ns (Fig. 6A). The motions mainly cause destabilization of the interface between the second and third AR thereby affecting the distance and relative orientation of ARs 1-2 relative to repeats 3-5 (Fig. 6B, C).

Subproject B

Although present feasible MD simulations in the ns-time range are still too short to monitor complete unfolding, the observed destabilization of the interface offers a structural explanation for the unfolding of the first and second AR on longer time scales, which has been experimentally detected for the phosphorylation-mimicking mutants.

Ubiquitination of p19

The introduction of a negative charge, localized between AR 2 and 3, seems enough to destabilize the native state *in vitro* in such a manner, that only the intermediate state with folded AR 3-5 gets populated at 37 °C. The modification of S76 by a negative charge occurs in the cell *via* phosphorylation, and raises the question about their functional role. Local unfolding of AR 1 and 2 after phosphorylation could allow the ubiquitin ligase to access lysine 62 more easily and finally target p19 to the proteasome.

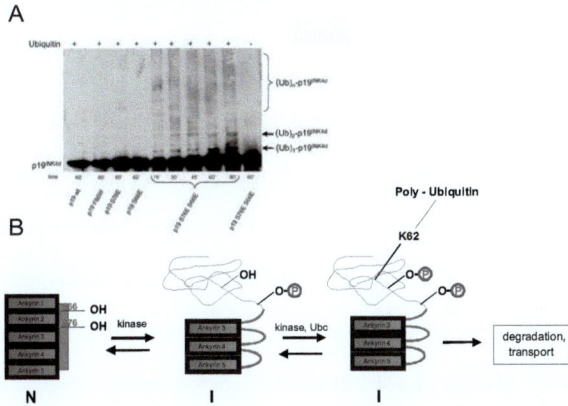

Fig. 7. (A) Ubiquitination of p19 *in vitro* by HeLa cell lysates requires at least two phosphorylation sites. The time course, mutants and positions of unmodified and ubiquitinated p19 are indicated. Reaction mix was resolved by SDS-PAGE (4-20%), and visualized by autoradiography using specific p19 antibodies. (B) Simplified phosphorylation model for p19: Phosphorylation of S76 leads to unfolding of AR 1 and 2 but a second phosphorylation at S66 is necessary for efficient ubiquitination and subsequent proteasomal degradation.

Because the specific ubiquitin ligase is not known, we tried to ubiquitinate p19 by using HeLa cell lysate extracts to test this hypothesis. Ubiquitination assays were negative for p19 wild type, while p19 S66E/S76E showed the strongest ubiquitination signal *in vitro* (Fig. 7A). Although mutation S66E has a minor effect on the stability, it seems to be important for ubiquitination. Single mutants containing either a glutamate at position 66 or 76 were hardly targeted by ubiquitination (Fig. 7A).

Thus, it is not surprising that mainly double phosphorylated p19 molecules were found in cell lines. Hence, phosphorylation of S76 leads to a local unfolding of p19, but a second phosphorylation is necessary for ubiquitination (Fig. 7B). These findings are in line with a suggested mechanism for phosphorylation targeted ubiquitination in eukaryotes (35). Ubiquitination of cyclin E for example strongly depends on posttranslational multisite phosphorylation, which induces binding to the SCF^{Fbw7} ubiquitin ligase complex (36, 37).

Conclusions

The present study combines experimental and computational techniques to study the impact of posttranslational phosphorylation on stability, folding, and function of CDK inhibitor p19. Substitution of serine 76 by a glutamate residue at the earlier identified phosphorylation site strongly reduced the stability of the native but not the intermediate state. Therefore, the intermediate state became significantly populated under equilibrium conditions. By raising the temperature to 37 °C, all molecules lost their native conformation allowing a detailed analysis of the intermediate state. Amide protons of AR 3-5 showed native chemical shift at body temperature, whereas all cross-peaks for AR 1-2 vanished. As earlier proposed (22), AR 3-5 host a scaffold function for the less stable repeats 1-2. MD simulations highlighted the molecular origin of the reduced stability. Phosphorylation of S76 strongly destabilizes the interface between AR 2 and 3, mainly by disturbing the hydrogen bonding network of adjacent residues. *In vitro* ubiquitination assays suggest a link of phosphorylated p19 and ubiquitination. While S76 phosphorylation results in an overall destabilization of the native molecule, S66 phosphorylation does not significantly affect the overall stability, but is nevertheless required for ubiquitination – most probably by changing the charge pattern in the immediate vicinity of the ubiquitin ligase binding site.

Table 1: Thermodynamic data of urea and temperature induced unfolding of p19^{INK4d} and variants

temperature	method	p19^{INK4d} variant	ΔG^0_{NU} (kJ/mol)^a	ΔG^0_{NI} (kJ/mol)	ΔG^0_{IU} (kJ/mol)	m_{NU} (kJ mol^{-1} M^{-1})^b	m_{NI} (kJ mol^{-1} M^{-1})	m_{IU} (kJ mol^{-1} M^{-1})
15 °C	equilibrium	pseudo wt^c	26.5 ± 0.9			8.48 ± 0.6		
15 °C	'I assay'	pseudo wt^c	30.6 ± 3.4	19.1 ± 1.2	11.5 ± 2.2	10.2 ± 1.4	5.5 ± 1.0	4.7 ± 0.4
15 °C	equilibrium	AR 3-5^c	9.62 ± 1.6			5.03 ± 1.2		
15 °C	kinetics	AR 3-5^c	7.97 ± 0.5			4.13 ± 0.4		
15 °C	equilibrium	S76E	14.5 ± 2.9	5.6 ± 1.6	8.9 ± 1.3	10.7 ± 1.6	5.2 ± 0.8	5.5 ± 0.8
37 °C	equilibrium	S76E			6.1 ± 1.2			4.6 ± 0.7
15 °C	equilibrium	S76E/S66E	13.1 ± 2.6	5.2 ± 1.5	7.9 ± 1.1	10.5 ± 1.2	5.4 ± 0.6	5.1 ± 0.6
15 °C	kinetics	S76E/S66E	11.3 ± 3.3	5.8 ± 1.3	5.5 ± 2.0	11.6 ± 1.9	5.9 ± 1.0	5.7 ± 0.9
37 °C	equilibrium	S76E/S66E			4.1 ± 1.8			4.3 ± 1.0
37 °C	kinetics	S76E/S66E^d			2.2 ± 1.3			4.5 ± 0.4
15 °C	equilibrium	S66E	25.6 ± 1.2			9.8 ± 0.8		
15 °C	equilibrium	S76A	24.9 ± 1.7			8.2 ± 0.5		

$^a \Delta G^0_{NU} = \Delta G^0_{NI} + \Delta G^0_{IU}$

$^b m_{NU} = m_{NI} + m_{IU}$

^{c} data taken from ref. (22)

$^d \Delta G^0_{IU}$ was underestimated due to negligence of the proline phase

Materials and Methods

Protein expression and purification

P19^{INK4d} and mutants were expressed and purified as described with minor modifications (38). The proteins were purified from soluble material, except mutants carrying the S76E mutation. These mutants were refolded from washed and urea solubilized inclusions bodies and purified to homogeneity. Correct folding state was confirmed by CD and NMR measurements as well as binding assays with CDK6. Mutations were introduced by using the Stratagene Qikchange

Kit (Stratagene, La Jolla, CA). Ubiquitination assays were performed with the S-100 *HeLa* Conjugation Kit (Boston Biochem; Cat. K-915) according to manufacturer instructions. Exogenously added p19 was detected by western blots using a monoclonal antibody.

Fluorescence spectroscopy

Urea was purchased from Gerbu and all other chemicals from Merck. Experiments were performed at 15 °C and/or 37 °C in 20 mM sodium-phosphate (pH 7.4). Urea transitions monitored by fluorescence were recorded with a JASCO FP6500 fluorescence spectrometer with an excitation wavelength of 280 nm and an emission wavelength between 300-420 nm at a protein concentration of 1-3 µM. The experimental data were analyzed according to a two- or three-state model (39) by non-linear least-squares fit with proportional weighting to obtain the Gibbs free energy of denaturation ΔG as a function of the urea concentration.

Kinetic mixing experiments

Fast kinetic experiments were performed using an Applied Photophysics SX-20MV stopped-flow instrument at 15 °C or 37 °C. An excitation wavelength of 280 nm was used and emission was monitored at wavelengths above 305 nm using cut-off filters. Unfolding experiments were performed by mixing protein in 20 mM sodium-phosphate (pH 7.4) with six or ten volumes of urea containing the same buffer. Refolding was initiated by 11- or 6-fold dilution of unfolded protein (4-6 M urea). Final protein concentration was 1-3 µM. Data collected from at least 4-8 scans were averaged and fitted using GraFit 5 (Erithacus Software). Unfolding traces were analyzed by single or double exponential functions, refolding traces by double or triple exponential functions. Chevron plots were analyzed with the program Scientist (MicroMath) according to two-state model or the analytical solutions of a three-state model (40, 41).

NMR measurements

NMR spectra were acquired with a Bruker Avance 800 spectrometer equipped with a cryoprobe in 20 mM sodium phosphate buffer, pH 7.4, containing 10 % ^2H_2O. ^{15}N-TROSY HSQC spectra of a 1 mM p19 wild type and a 0.4 mM p19 S76E sample were recorded between 15 °C and 50 °C. The same experiments were performed with a 1 mM ^{15}N-labeled sample of an unfolded peptide, which was used as a reference. All samples were cooled down to 25° C afterwards to check reversibility. Unfolded signals of the p19 mutant and the peptide could not be observed at higher temperatures due to the fast exchange of amide protons with water. Spectra were processed with NMRpipe (42) and analyzed with NMRView (43). Only signals of the native population were used for analysis.

MD simulations

For the MD analysis four model systems representing unphosphorylated, S66-phosphorylated, S76-phosphorylated, and S66+S76-phosphorylated p19 were generated based on the structure of PDB entry 1bd8 (9). The terminal residues (R7, M162) that were resolved in the crystal structure were blocked by acetyl- and N-methyl-groups. Missing atoms were added with the LEAP module of AMBER 7 (44). Neutralization and solvation of the systems was performed as described in (45).

Minimizations and MD simulations were carried out with the PMEMD module of AMBER 9 (46) using the parm99 force field (47, 48) augmented by parameters for unprotonated phospho-serine (49). After minimization of the solvent, the whole systems were consecutively subjected to 250 steps steepest descent and 19,750 steps conjugate gradient minimization. Before the MD production was started, the systems were equilibrated by a 100 ps MD run, in which the temperature was increased form 50 to 310 K, and an 80 ps MD simulation, in which the density was adjusted to 1 g cm^{-3}. MD simulations were performed using an 8.0 Å cutoff and a time step of 2 fs for the integration of the equations of motion. All other MD settings were identical to those specified in (45). Collection of simulation data took place in time intervals of 2 ps along the trajectories. The data were analyzed with the programs X-PLOR (50), PTRAJ (44), and VMD (51).

Acknowledgements

We thank P. Rösch for NMR spectrometer time at 800MHz, Christine Magg for help with western blots and the members of our group for helpful discussions. This research was supported by grants from the Deutsche Forschungsgemeinschaft (Ba 1821/3-1, GRK 1026, SFB 473) and the excellence initiative of the state Sachsen-Anhalt.

References

1. Pines J (1996) Cell cycle: reaching for a role for the Cks proteins. *Curr. Biol.* 6:1399-1402.
2. Morgan DO (1995) Principles of CDK regulation. *Nature* 374:131-134.
3. Bartek J, Bartkova J, Lukas J (1996) The retinoblastoma protein pathway and the restriction point. *Curr Opin Cell Biol* 8:805-14.
4. Sherr CJ (1996) Cancer cell cycles. *Science* 274:1672-1677.
5. Lees E (1995) Cyclin dependent kinase regulation. *Curr Opin Cell Biol* 7:773-80.
6. Sherr CJ, Roberts JM (1999) CDK inhibitors: positive and negative regulators of G1-phase progression. *Genes. Dev.* 13:1501-1512.
7. Harper JW, Elledge SJ (1996) Cdk inhibitors in development and cancer. *Curr. Opin. Genet. Dev.* 6:56-64.
8. Bartek J, Bartkova J, Lukas J (1997) The retinoblastoma protein pathway in cell cycle control and cancer. *Exp Cell Res* 237:1-6.
9. Baumgartner R, Fernandez-Catalan C, Winoto A, Huber R, Engh RA, Holak TA (1998) Structure of human cyclin-dependent kinase inhibitor p19INK4d: comparison to known ankyrin-repeat-containing structures and implications for the dysfunction of tumor suppressor p16INK4a. *Structure* 6:1279-1290.
10. Luh FY, et al. (1997) Structure of the cyclin-dependent kinase inhibitor p19Ink4d. *Nature* 389:999-1003.
11. Byeon IJ, Li J, Ericson K, Selby TL, Tevelev A, Kim HJ, O'Maille P, Tsai MD (1998) Tumor suppressor p16INK4A: determination of solution structure and analyses of its interaction with cyclin-dependent kinase 4. *Mol Cell* 1:421-31.
12. Li J, Byeon IJ, Ericson K, Poi MJ, O'Maille P, Selby T, Tsai MD (1999) Tumor suppressor INK4: determination of the solution structure of p18INK4C and demonstration of the functional significance of loops in p18INK4C and p16INK4A. *Biochemistry* 38:2930-2940.
13. Yuan C, Selby TL, Li J, Byeon IJ, Tsai MD (2000) Tumor suppressor INK4: refinement of p16INK4A structure and determination of p15INK4B structure by comparative modeling and NMR data. *Protein Sci.* 9:1120-1128.
14. Serrano M (1997) The tumor suppressor protein p16INK4a. *Exp Cell Res* 237:7-13.
15. Ruas M, Peters G (1998) The p16INK4a/CDKN2A tumor suppressor and its relatives. *Biochim Biophys Acta* 1378:F115-77.
16. Drexler HG (1998) Review of alterations of the cyclin-dependent kinase inhibitor INK4 family genes p15, p16, p18 and p19 in human leukemia-lymphoma cells. *Leukemia* 12:845-59.
17. Bartkova J, Thullberg M, Rajpert-De Meyts E, Skakkebaek NE, Bartek J (2000) Lack of p19INK4d in human testicular germ-cell tumours contrasts with high expression during normal spermatogenesis. *Oncogene* 19:4146-50.

18. Thullberg M, Bartek J, Lukas J (2000) Ubiquitin/proteasome-mediated degradation of p19INK4d determines its periodic expression during the cell cycle. *Oncogene* 19:2870-2876.
19. Thullberg M, Bartkova J, Khan S, Hansen K, Ronnstrand L, Lukas J, Strauss M, Bartek J (2000) Distinct versus redundant properties among members of the INK4 family of cyclin-dependent kinase inhibitors. *FEBS Lett.* 470:161-166.
20. Scassa ME, Marazita MC, Ceruti JM, Carcagno AL, Sirkin PF, Gonzzalez-Cid M, Pignataro OP, Canepa ET (2007) Cell cycle inhibitor, p19INK4d, promotes cell survival and decreasese chromosomal aberrations after genotoxic insult due to enhanced DNA repair. *DNA repair*:[Epub ahead of print].
21. Ceruti JM, Scassa ME, Flo JM, Varone CL, Canepa ET (2005) Induction of p19INK4d in response to ultraviolet light improves DNA repair and confers resistance to apoptosis in neuroblastoma cells. *Oncogene* 24:4065-4080.
22. Löw C, Weininger U, Zeeb M, Zhang W, Laue ED, Schmid FX, Balbach J (2007) Folding mechanism of an ankyrin repeat protein: scaffold and active site formation of human CDK inhibitor p19(INK4d). *J Mol Biol* 373:219-31.
23. Zeeb M, Rösner H, Zeslawski W, Canet D, Holak TA, Balbach J (2002) Protein folding and stability of human CDK inhibitor p19INK4d. *J. Mol. Biol.* 315:447-457.
24. Tang KS, Fersht AR, Itzhaki LS (2003) Sequential unfolding of ankyrin repeats in tumor suppressor p16. *Structure* 11:67-73.
25. Tang KS, Guralnick BJ, Wang WK, Fersht AR, Itzhaki LS (1999) Stability and folding of the tumour suppressor protein p16. *J. Mol. Biol.* 285:1869-1886.
26. Yuan C, Li J, Selby TL, Byeon IJ, Tsai MD (1999) Tumor suppressor INK4: comparisons of conformational properties between p16(INK4A) and p18(INK4C). *J Mol Biol* 294:201-11.
27. Hunter T (2000) Signaling--2000 and beyond. *Cell* 100:113-27.
28. Johnson LN, Barford D (1993) The effects of phosphorylation on the structure and function of proteins. *Annu Rev Biophys Biomol Struct* 22:199-232.
29. Kreegipuu A, Blom N, Brunak S (1999) PhosphoBase, a database of phosphorylation sites: release 2.0. *Nucleic Acids Res* 27:237-9.
30. Steen H, Jebanathirajah JA, Springer M, Kirschner MW (2005) Stable isotope-free relative and absolute quantitation of protein phosphorylation stoichiometry by MS. *Proc Natl Acad Sci U S A* 102:3948-53.
31. Hupp TR, Lane DP (1995) Two distinct signaling pathways activate the latent DNA binding function of p53 in a casein kinase II-independent manner. *J Biol Chem* 270:18165-74.
32. Li M, Stukenberg PT, Brautigan DL (2008) Binding of phosphatase inhibitor-2 to prolyl isomerase Pin1 modifies specificity for mitotic phosphoproteins. *Biochemistry* 47:292-300.
33. Park KS, Mohapatra DP, Misonou H, Trimmer JS (2006) Graded regulation of the Kv2.1 potassium channel by variable phosphorylation. *Science* 313:976-9.
34. Mello CC, Bradley CM, Tripp KW, Barrick D (2005) Experimental characterization of the folding kinetics of the notch ankyrin domain. *J. Mol. Biol.* 352:266-281.
35. Feldman RM, Correll CC, Kaplan KB, Deshaies RJ (1997) A complex of Cdc4p, Skp1p, and Cdc53p/cullin catalyzes ubiquitination of the phosphorylated CDK inhibitor Sic1p. *Cell* 91:221-30.
36. Koepp DM, Schaefer LK, Ye X, Keyomarsi K, Chu C, Harper JW, Elledge SJ (2001) Phosphorylation-dependent ubiquitination of cyclin E by the SCFFbw7 ubiquitin ligase. *Science* 294:173-7.
37. Hao B, Oehlmann S, Sowa ME, Harper JW, Pavletich NP (2007) Structure of a Fbw7-Skp1-cyclin E complex: multisite-phosphorylated substrate recognition by SCF ubiquitin ligases. *Mol Cell* 26:131-43.
38. Kalus W, Baumgartner R, Renner C, Noegel A, Chan FK, Winoto A, Holak TA (1997) NMR structural characterization of the CDK inhibitor p19INK4d. *FEBS Lett* 401:127-132.
39. Hecky J, Müller KM (2005) Structural perturbation and compensation by directed evolution at physiological temperature leads to thermostabilization of beta-lactamase. *Biochemistry* 44:12640-12654.
40. Ikai A, Tanford C (1973) Kinetics of unfolding and refolding of proteins. I. Mathematical analysis. *J. Mol. Biol.* 73:145-163.
41. Bachmann A, Kiefhaber T (2005) in *Protein Folding Handbook*, eds. Buchner, J. & Kiefhaber, T. (Wiley-VCH, Weinheim), pp 379-406.
42. Delaglio F, Grzesiek S, Vuister GW, Zhu G, Pfeifer J, Bax A (1995) NMRPipe: a multidimensional spectral processing system based on UNIX pipes. *J. Biomol. NMR* 6:277-93.
43. Johnson BA, Blevins RA (1994) NMRView: A computer program for visualization and analysis of NMR data. *Journal of Biomolecular NMR* 4:603-614.
44. Case DA, et al. (2002) AMBER 7. (University of California, San Francisco, USA).
45. Homeyer N, Essigke T, Meiselbach H, Ullmann GM, Sticht H (2007) Effect of HPr phosphorylation on structure, dynamics, and interactions in the course of transcriptional control. *J Mol Model* 13:431-44.
46. Case DA, et al. (2006) AMBER 9. (University of California, San Francisco).
47. Cornell WD, et al. (1995) A Second Generation Force Field for the Simulation of Proteins, Nucleic Acids and Organic Molecules. *J. Am. Chem. Soc.* 117:5179-97.

48. Cheatham TE, 3rd, Cieplak P, Kollman PA (1999) A modified version of the Cornell et al. force field with improved sugar pucker phases and helical repeat. *J. Biomol. Struct. Dyn.* 16:845-62.
49. Homeyer N, Horn AHC, Lanig H, Sticht H (2006) AMBER force-field parameters for phosphorylated amino acids in different protonation states: phosphoserine, phosphothreonine, phosphotyrosine, and phosphohistidine. *J. Mol. Model.* 12:281-9.
50. Simonson T, Brunger AT (1992) Thermodynamics of Protein-Peptide Interactions in the Ribonuclease-S System Studied by Molecular Dynamics and Free Energy Calculations. *Biochemistry* 31:8661-8674.
51. Humphrey W, Dalke A, Schulten K (1996) VMD: visual molecular dynamics. *J Mol Graph* 14:33-38.
52. Koradi R, Billeter M, Wüthrich K (1996) MOLMOL: a program for display and analysis of macromolecular structures. *J Mol Graph* 14:51-55.

7.3 Subproject C

Subproject C

Subproject C

Structural insights into an equilibrium folding intermediate of an archaeal ankyrin repeat protein

Christian Löw*, Ulrich Weininger*, Piotr Neumann[†], Mirjam Klepsch[‡], Hauke Lilie[§], Milton T. Stubbs[†¶], and Jochen Balbach*[¶‖]

*Institut für Physik, Biophysik, [†]Institut für Biochemie und Biotechnologie, Physikalische Biochemie, [§]Institut für Biochemie/Biotechnologie, Biotechnologie, and [¶]Mitteldeutsches Zentrum für Struktur und Dynamik der Proteine (MZP), Martin-Luther-Universität Halle-Wittenberg, D-06120 Halle (Saale), Germany; and [‡]Department of Biochemistry and Biophysics, Center for Biomembrane Research, Stockholm University, SE-10691 Stockholm, Sweden

Edited by Robert T. Sauer, Massachusetts Institute of Technology, Cambridge, MA, and approved January 9, 2008 (received for review November 9, 2007)

Repeat proteins are widespread in nature, with many of them functioning as binding molecules in protein–protein recognition. Their simple structural architecture is used in biotechnology for generating proteins with high affinities to target proteins. Recent folding studies of ankyrin repeat (AR) proteins revealed a new mechanism of protein folding. The formation of an intermediate state is rate limiting in the folding reaction, suggesting a scaffold function of this transient state for intrinsically less stable ARs. To investigate a possible common mechanism of AR folding, we studied the structure and folding of a new thermophilic AR protein (tANK) identified in the archaeon *Thermoplasma volcanium*. The x-ray structure of the evolutionary much older tANK revealed high homology to the human CDK inhibitor p19[INK4d], whose sequence was used for homology search. As for p19[INK4d], equilibrium and kinetic folding analyses classify tANK to the family of sequential three-state folding proteins, with an unusual fast equilibrium between native and intermediate state. Under equilibrium conditions, the intermediate can be populated to >90%, allowing characterization on a residue-by-residue level using NMR spectroscopy. These data clearly show that the three C-terminal ARs are natively folded in the intermediate state, whereas native cross-peaks for the rest of the molecule are missing. Therefore, the formation of a stable folding unit consisting of three ARs is the necessary rate-limiting step before AR 1 and 2 can assemble to form the native state.

folding kinetics | protein folding | NMR | protein structure | *Thermoplasma volcanium*

Ankyrin repeat (AR) proteins are ubiquitous and involved in numerous fundamental physiological processes (1). A common feature of repeat proteins from all families is their modular architecture of homologous structural elements forming a scaffold for specific and tight molecular interactions. This property has been applied in biotechnology to generate AR proteins with high affinities for target proteins (2). The AR consists of 33 amino acids that form a loop and a β-turn followed by two antiparallel α-helices connected by a tight turn. Up to 29 repeats can be found in a single protein, but usually four to six repeats stack onto each other to form an elongated structure with a continuous hydrophobic core and a large solvent accessible surface (3–6). Unlike the packing of globular protein domains, the linear arrangement of the repeat modules in AR proteins implies that local, regularly repeating packing interactions are very important and may dominate the thermodynamic stability and the folding mechanism. AR proteins are therefore expected to fold in a fast, modular, multistate reaction controlled by short-range interactions. Interestingly, folding of naturally occurring AR proteins is much slower than expected from the low contact order and shows almost exclusively two-state behavior under equilibrium conditions (7–16). However, kinetic and equilibrium analysis of the folding of CDK inhibitor p19[INK4d] (17, 18) and Notch ankyrin domain (19), revealed a surprising folding mechanism including the formation of an intermediate state as rate limiting step. Partially folded intermediate states found on the folding pathway of small globular domains usually form much faster than the rate limiting step of folding. Thus, the intermediate state of an AR protein may act as a scaffold, requiring initial folding before zipping up of the less stable repeats in a fast reaction to the native state.

Folding studies on naturally occuring AR proteins have until now focused only on eukaryotic proteins. To test the validity of a possible common mechanism of AR folding, we performed a Blast search with the p19[INK4d] sequence as template on evolutionary much older archaeal organisms. A new protein of similar length and with <25% sequence identity to p19[INK4d] was identified in *Thermoplasma volcanium* (20). The herein determined structure by x-ray crystallography confirmed that this archaeal AR protein (tANK) folds into five sequentially arranged ARs with an additional helix at the N terminus. Equilibrium and kinetic folding analyses of this protein by fluorescence and CD spectroscopy revealed a sequential three-state folding mechanism with the expected unusual fast equilibrium between the native and intermediate state. Compared with p19[INK4d] and the Notch ankyrin domain, the intermediate state of tANK can be populated to 90% at equilibrium, making high-resolution studies possible. GdmCl induced equilibrium unfolding transitions monitored by NMR showed that the amide protons of AR 3–5 in the intermediate still resonate at native chemical shifts whereas the N-terminal AR are mainly unfolded. Limited proteolysis data confirmed AR 3–5 as the most stable part of the protein.

Results and Discussion

X-Ray Structure of tANK. To compare protein folding data derived from human AR proteins with thermophilic AR proteins, we performed a Blast search against the archaean database using the human p19[INK4d] sequence as template. A putative AR protein (tANK) was identified in *Thermoplasma volcanium* sharing <25% sequence identity with the human p19[INK4d] protein [see supporting information (SI) Fig. 7]. The top hits in a second Blast search with the sequence of tANK in the nonredundant protein database only comprised archaeal homologs. Therefore, a horizontal gene transfer seems to be unlikely. The crystal structure of this protein was solved to 1.65 Å resolution, confirming the expected five-membered AR fold.

Author contributions: C.L. and J.B. designed research; C.L., U.W., P.N., and H.L. performed research; C.L., U.W., P.N., M.K., and H.L. analyzed data; and C.L., M.T.S., and J.B. wrote the paper.

The authors declare no conflict of interest.

This article is a PNAS Direct Submission.

Data deposition: The data reported in this paper have been deposited in the Protein Data Bank, www.pdb.org (PDB ID code 2RFM).

[‖]To whom correspondence should be addressed. E-mail: jochen.balbach@physik.uni-halle.de.

This article contains supporting information online at www.pnas.org/cgi/content/full/0710657105/DC1.

© 2008 by The National Academy of Sciences of the USA

Subproject C

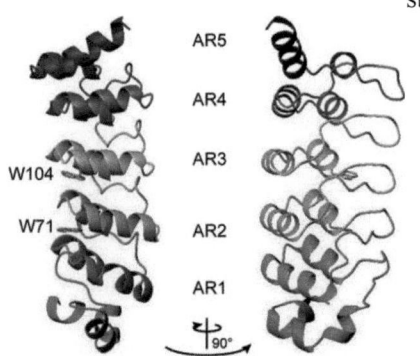

Fig. 1. Schematic representation of the structure of the thermophilic ankyrin repeat protein tANK (Protein Data Base ID code 2RFM). Five ARs (AR1–AR5), each comprising a loop, a β-turn, and two sequential α-helices form the elongated structure, extended by an α-helical N terminus are shown. Side chains of the wild-type fluorescence probes Trp-71 and Trp-104 are indicated as sticks. The figure was created by using MOLMOL (34).

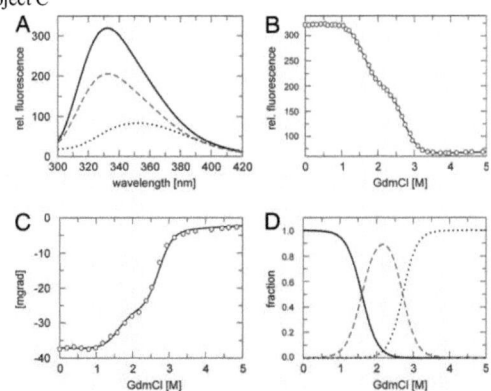

Fig. 2. GdmCl-induced unfolding of tANK monitored by fluorescence and CD spectroscopy. (A) Fluorescence spectra of tANK at 0 M (black line), 2 M (broken gray line) and 5 M (dotted black line) GdmCl after excitation at 280 nm. GdmCl induced unfolding transitions monitored by fluorescence at 335 nm (B) and CD at 222.6 nm (C). Solid lines in B and C represent the least square fit of a three-state model (see Materials and Methods for details). (D) Calculated equilibrium populations of the native N (black line), intermediate I (broken gray line), and unfolded state U (dotted black line) according to the global analysis of fluorescence and CD equilibrium data.

Structural refinement procedures and statistics are given in the supporting information. Compared with p19^{INK4d} (21, 22), the thermophilic protein harbours an extension of 23 aa at the N terminus which forms an additional helix (Fig. 1). The backbone rmsd for AR 3–5 between the mesophilic and thermophilic protein was <1.5 Å, indicating the high conservation of this structure element in evolution.

GdmCl-Induced Unfolding Involves the Formation of a Partially Structured Intermediate. The stability of tANK was monitored by fluorescence- and CD spectroscopy in the presence of various amounts of GdmCl. Trp-71 and Trp-104 located in AR 2 and 3 (Fig. 1) proved excellent probes to follow the transition curve. Upon unfolding, the fluorescence of the native protein N is strongly quenched and the maximum of the spectrum is shifted to higher wavelength (333 nm → 355 nm). At medium concentrations of GdmCl, however, an intermediate state I gets populated, which shows a quenched maximum still at 333 nm (Fig. 2A). To follow the α-helical content of the protein upon GdmCl unfolding, far-UV CD at 222 nm was used as a second probe. At medium concentrations of GdmCl (2 M) one third of the native CD signal is lost, indicating the unfolding of some secondary structure elements (Fig. 2C). A combined analysis of GdmCl-induced unfolding curves monitored by fluorescence (Fig. 2B) and CD according to a three-state model revealed a global stability of ΔG_u = 52.6 ± 1.8 kJ/mol for tANK; 18.5 ± 1.0 kJ/mol count for the N to I transition, whereas the intermediate state has a stability of 34.1 ± 1.5 kJ/mol relative to U. Resulting m values are 11.6 ± 0.7 kJ·mol^{-1} M^{-1} and 12.6 ± 0.6 kJ·mol^{-1} M^{-1} for the first and second transitions, respectively. Calculated populations of N, I, and the unfolded state (U) according to these biophysical parameters show that I is populated to an extent of 90% under equilibrium conditions at ≈2.1 M GdmCl (Fig. 2D).

Analytical ultracentrifugation of tANK indicates that the intermediate state is monomeric: the sedimentation equilibration at 1.95 M GdmCl gave an M_r of 22.7 ± 1.6 kDa, which is consistent with the mass of one polypeptide chain (see SI Fig. 8). However, at this GdmCl concentration, a 1 mM protein sample forms fibrils after several days (see SI Fig. 11). This is consistent with the idea that populated but destabilized folding intermediates are prone to the formation of ordered aggregates (23).

Folding Kinetics. Unfolding and refolding kinetics of tANK were measured by stopped-flow fluorescence spectroscopy. Unfolding under fully denaturing conditions (>3 M GdmCl) is fast and best described by a biphasic process with rate constants that differ by at least a factor of 20 (Fig. 3A). Each unfolding reaction contributes to 50% of the whole unfolding amplitude (Fig. 3C). Below 3 M GdmCl, however, the amplitude for the slow unfolding reaction decreases faster compared with the amplitude for the fast unfolding reaction, indicating the population of the intermediate state. At 2 M GdmCl, for example, where I is the dominant species, the fast unfolding phase accounts for 90% of the amplitude. The refolding reaction starting from fully unfolded protein is best described by three exponential functions with folding rates that differ by more than a factor of 10 (Fig. 3B). The fast reaction adds to >85% of the amplitude, whereas the two slow reactions account for <10% each (Fig. 3D). Nevertheless, at low GdmCl concentrations, no fast refolding phase in the range between 100 s^{-1} and 1,000 s^{-1} was observed in this experiment. Note that the entire refolding amplitude is detectable during the refolding reaction, evident from the start and end point analysis of the kinetics (Fig. 3F). Because of the latter observation, burst phase intermediates or problems in reversibility can be excluded. These findings can be explained by the sequential folding mechanism U ⇌ I ⇌ N already observed for p19^{INK4d} and the Notch ankyrin domain. The formation of the intermediate state is rate limiting during the refolding reaction. Kinetic experiments using single mixing can only directly monitor reactions that occur before the rate limiting step if the rates differ by >5-fold. This explains the absence of a very fast refolding phase between 0 and 1 M GdmCl (Fig. 3E). To confirm this assumption, unfolding and refolding reactions were initiated from the intermediate state. Protein was incubated at 1.7 M GdmCl (45% native, 55% I, 0% U) and refolded to native conditions between 0.4 and 1.6 M GdmCl. All refolding kinetics were very fast and followed a single exponential function (Fig. 3B Inset). These rates filled the missing gap of the Chevron plot (open gray symbols in Fig. 3E) and thereby assigned the fast

Subproject C

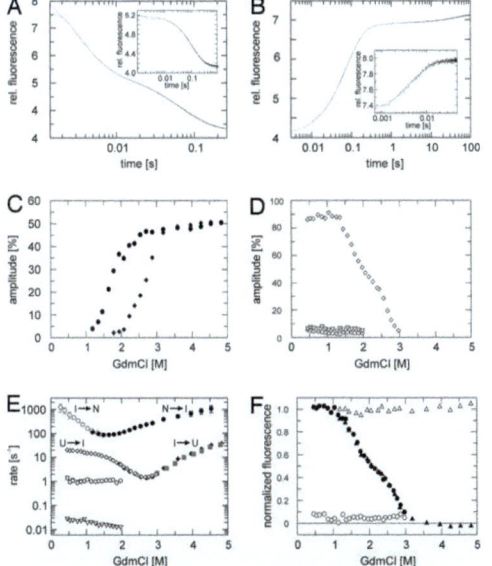

Fig. 3. Single mixing unfolding and refolding kinetics of tANK detected by stopped flow fluorescence. (*A* and *B*) Experimental data are plotted in black and fits in gray. (*A*) Unfolding was initiated by a rapid change from 0 M to 4 M GdmCl and can be best fitted by a single exponential function. (*B*) Refolding was initiated by rapid dilution from 5 M to 0.9 M GdmCl and follows a sum of three exponentials. Refolding (*A* Inset) and unfolding traces (*B* Inset) starting from GdmCl concentrations where the intermediate is highly populated (1.7 M GdmCl for refolding and 2.6 M GdmCl for unfolding) can be best described by a single exponential function. Amplitudes of refolding (*C*) and unfolding (*D*) phases, calculated as a percentage of the total fluorescence change between the native and unfolded state: filled diamond, slow phase; filled circle, fast phase of unfolding; the amplitude of the refolding kinetics is dominated by one phase (open diamonds) and two minor phases (open hexagon, slow; open inverted triangle, very slow) with <10% of the whole amplitude are detectable. (*E*) GdmCl dependence of apparent folding rates of tANK monitored at 15°C, pH 7.4. Filled symbols indicate unfolding experiments, open symbols indicate refolding experiments. Gray symbols represent folding rates that result from unfolding and refolding kinetics starting from the intermediate state. (*F*) Start and end point analysis of the kinetic experiments. End points of unfolding (filled triangle) and refolding (filled circle) reactions follow fluorescence equilibrium data (Fig. 2*B*). Start point (open circle and triangle) analysis reveal that there is no obvious burst-phase observable.

folding reaction to the transition between the native and intermediate state. Below 0.5 M GdmCl, the rates exceed the limits of conventional stopped-flow techniques.

Unfolding reactions at high GdmCl concentration with protein preincubated at 2.6 M GdmCl (0% native, 55% I, 45% U) follow a single exponential decay (Fig. 3*A* Inset). These unfolding rates (closed gray symbols in Fig. 3*E*) match with slow unfolding rates observed in unfolding experiments starting from the native state. Therefore, we assign these rates to the slow reaction between the intermediate and the unfolded state. These data clearly confirm the early observations seen for p19[INK4d] and the Notch ankyrin domain, namely that the formation of the intermediate state is indeed the rate limiting step in the folding reaction.

Although logarithms of the fastest un-/folding rates show a linear dependence of the GdmCl concentration, a pronounced kink in the unfolding and refolding limb of the U ↔ I transition is visible. These "roll overs" are not caused by kinetic coupling of the observed folding rates, because the refolding and unfolding rates differ by more than a factor of 10. Furthermore, unfolding rates derived from unfolding reactions initiated from the intermediate state also result in a downward curvature of the unfolding limb. Curvatures in the refolding and unfolding limbs of the Chevron plot have been observed for various proteins (24, 25). These nonlinearities are often interpreted in terms of a broad energy barrier, where the addition of denaturant can cause a movement of the transition state which results in kinetic anomalies (25). Alternatively, the existence of additional high energy intermediates in unfolding and refolding reactions can explain these observations (26).

In contrast to the fast reactions, the two minor phases detected in refolding reactions starting from the unfolded state do not show a significant dependence on GdmCl concentration (Fig. 3*E*). None of these two folding reactions is detectable when refolding is initiated from the intermediate state, indicating that these reactions originate from heterogeneity in the unfolded state. The temperature dependence of the slowest refolding rate yields in an activation energy of 78.5 ± 3 kJ/mol, typical for prolyl *cis/trans* isomerization reactions (27) (see SI Fig. 10). This assumption could be confirmed by a 5-fold acceleration of this folding phase in the presence of 10% of the prolyl isomerase SlyD from *Thermus thermophilus* (C.L., unpublished results, data not shown). It should be noted that all prolyl peptide bonds of the native state are in the *trans* conformation according to the crystal structure of tANK.

The second slow folding rate has a GdmCl independent time constant of 1 s at 15°C. Although the amplitude for this reaction is quite small (<5%), the refolding rate could be accurately determined due to the large change in fluorescence between the native and the unfolded state. The activation energy for this reaction was determined to 47.5 ± 3 kJ/mol (see SI Fig. 10). Compared with literature data (28), this suggests that the origin of this folding rate is caused by the isomerization of nonprolyl *cis* peptide bonds in the unfolded state.

GdmCl Folding Transition Monitored by NMR. As a result of the high population of the intermediate state under equilibrium conditions, it was possible to further characterize this state by NMR spectroscopy. To this end, >83% of the backbone amide protons of the native state were assigned using standard 3D experiments (see SI Fig. 9). The GdmCl induced transition of tANK was followed by a series of 19 2D ^{15}N-TROSY-HSQC spectra recorded at various GdmCl concentrations. Long incubation times of the NMR samples at medium concentrations of GdmCl resulted in fibril formation (see above). Thus for each datapoint a fresh sample was prepared in the transition region of the GdmCl transition. 66 out of 185 possible native amide cross-peaks could be followed during the entire transition without overlap of cross-peaks from I and U. Fig. 4 depicts two sections from these ^{15}N-TROSY-HSQCs, where cross-peaks of the native state disappear at low (e.g., G88) or at high (e.g., G121) GdmCl concentrations. Some cross-peaks appear only at intermediate GdmCl concentrations (e.g., Int1 and Int2). It should be noted that the selected sections are outside the range of cross-peaks of the unfolded state and that the chemical exchange between the three states is slow compared with chemical shift time scale at GdmCl concentrations, were U, I, and N are highly populated.

For quantitative analysis, the volume of each cross-peak present at 0 M GdmCl was plotted against the GdmCl concentration, resulting in 66 unfolding transition curves (examples are shown in Fig. 5*A*). They can be grouped into two classes, following the two transitions seen from fluorescence and CD

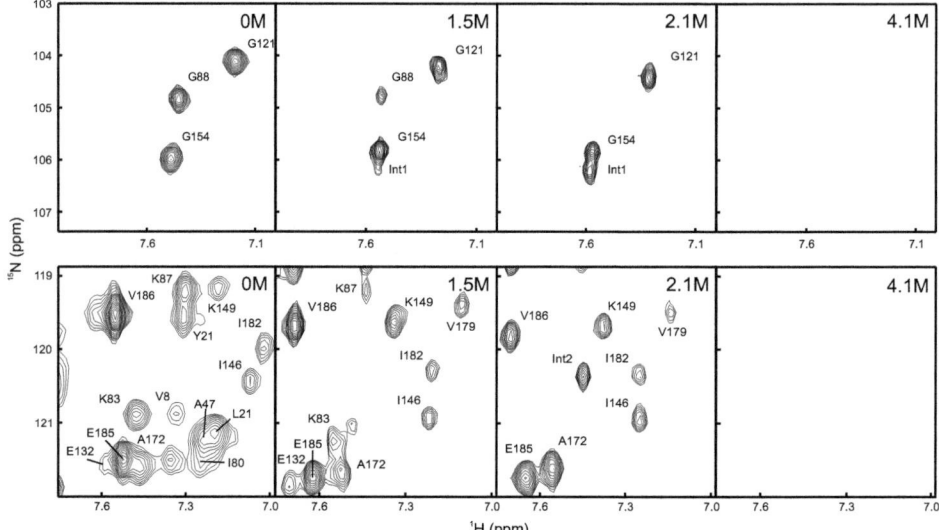

Fig. 4. Sections of ^{15}N-TROSY-HSQC spectra of tANK show the disappearance of native cross-peaks at low (e.g., G88) and high (e.g., G121) denaturation concentrations. Transiently appearing cross-peaks of the intermediate state are labeled with Int. Residues of AR 3–5 are still present at 2.1 M GdmCl, where the intermediate state is maximally populated, whereas signals of the N-terminal part are missing. This indicates that AR 3–5 remain folded in the intermediate state.

data (Fig. 2). Unfolding curves derived from N-terminal repeats show a transition midpoint of ≈1.6 M GdmCl (Fig. 5D). These cross-peaks with native chemical shifts vanished at 2.1 M GdmCl, where the intermediate state is maximally populated. In contrast, residues of AR 3–5 still show native chemical shifts at 2.1 M GdmCl and also unfold cooperatively with a transition midpoint of ≈2.6 M GdmCl (Fig. 5D). Detailed analysis allows assignment of the two transitions observed by optical methods to the respective residues in tANK. Amide protons of the first ARs follow the decay of the native population as derived from fluorescence and CD data. However, residues from AR 3–5 can be described by the sum of the native and intermediate population. This demonstrates that in the intermediate state, amide protons of AR 3–5 still show native chemical shifts whereas resonances of residues from AR 1 and 2 show nonnative chemical shifts. Moreover, the build-up of 40 unfolded cross-peaks could be followed over the entire GdmCl range (examples shown in Fig. 5C). Interestingly, they show a corresponding pattern, where some are already maximally populated in the intermediate state, suggesting some N-terminal residues sense a completely unfolded chemical environment. The larger fraction of unfolded peaks follows the decay of the intermediate state. Furthermore, 12 additional peaks could be directly assigned to the intermediate state far off random coil shifts (e.g., Int1 in Fig. 4), which arise during the first transition. They get fully populated at 2.1 M GdmCl and then decay at higher GdmCl concentrations. The course of additional intermediate signals with GdmCl concentration (Fig. 5B) agrees well with the intermediate population calculated from the fluorescence data (Fig. 2D). Transition curves resulting from the N-terminal helical extension (residues 1–24) of tANK show much lower midpoints compared with the rest of the molecule (Fig. 5D). These data match with limited proteolysis data (see SI Table 2), which showed a rapid degradation of the N-terminal 25 residues.

Longer incubation times, however, lead to a stable 10 kDa fragment as judged by SDS page, identified as AR 3–5 by mass spectroscopy. The proteolysis data therefore verify the graded thermodynamic stability of tANK found from GdmCl induced unfolding transitions.

Sequential Folding Mechanism of tANK. The biophysical data presented here suggest the simplified model for folding of tANK outlined in Fig. 6. Folding and unfolding of tANK is a sequential, discrete process via an on-pathway intermediate with a highly cooperative transition between the consecutive steps. The partially folded state contains folded AR 3–5. The N-terminal part of the intermediate might contain some residual secondary structure indicated by the CD detected unfolding transition (Fig. 2C) and the good dispersion of NMR chemical shifts from some residues of this region (Fig. 4). The formation of this intermediate state is rate limiting for the refolding reaction, which suggests a scaffold function for AR 3–5. Interestingly, sequences of AR 3–5 of tANK and p19^{INK4d} show a high homology to the consensus AR sequence (29, 30). Designed AR proteins based on this consensus sequence are known to be significantly more stable compared with naturally occurring AR proteins (29). Therefore we propose that ARs with the highest local stability (usually two to three repeats) form the initiation site of the folding reaction.

Materials and Methods

Gene Construction, Protein Expression and Purification. A living culture of *Thermoplasma volcanium* (DSM 4300) was purchased from DSMZ (Deutsche Sammlung von Mirkroorganismen und Zellkulturen GmbH, Braunschweig, Germany). The organism was grown in *Thermoplasma volcanium* medium (medium 398) under anaerobic conditions at 60°C for 2 weeks. Genomic DNA was prepared using the Wizard DNA Purification System (Promega). The gene for the thermophilic AR protein was amplified by using flanking primers and cloned into a pet28c vector. The gene sequence was confirmed by automated

Subproject C

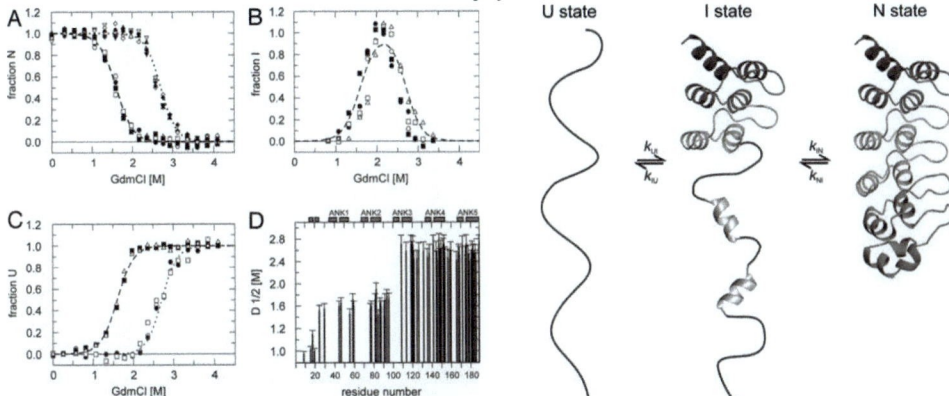

Fig. 5. GdmCl induced unfolding transitions of tANK monitored by NMR. ^{15}N-TROSY-HSQC spectra were recorded between 0 and 4.2 M GdmCl. (*A*) Normalized cross-peak volumes of backbone amides assigned to the native state at 0 M GdmCl. E45, D60, L78, G88, and V91 of AR 1–2 follow the decay of the native state population derived from the fluorescence and CD data (broken line). G109, E119, G142, L153, and A189 of AR 3–5 follow the sum of the native state and intermediate state population derived from the fluorescence and CD data (dotted lines). (*B*) Additional, transient cross-peaks which do not heavily overlap with peaks from the native or denatured state agree with the intermediate population (broken line) resulting from fluorescence and CD data. (*C*) The build-up of cross-peak volumes of the denatured state was monitored over the entire GdmCl range. Some cross-peaks are already maximal at intermediate GdmCl concentrations (2.1 M GdmCl), whereas the volumes of other cross-peaks increased with unfolding of the intermediate state. The data suggest that some residues experience a denatured like environment in the intermediate state. (*D*) Midpoints of denaturation profiles of 66 of 185 possible amide cross-peaks show that the two N-terminal AR are by 1 M GdmCl less stable compared with C-terminal three repeats.

Fig. 6. Simplified folding model of tANK. The protein folds via an on-pathway intermediate in which the two N-terminal repeats are unfolded and the three C-terminal repeats are natively folded. Additional NMR cross-peaks of the intermediate state, which do not show a random coil chemical shift, and CD data suggest that there is some residual secondary structure in AR 1 and 2.

DNA sequencing. Protein was expressed in *Escherichia coli* BL21 (DE3) and purified from soluble material. Cells were resuspended in IMAC binding buffer (50 mM Tris, 300 mM NaCl, 20 mM Imidazole, pH 8.0) and lysed by sonication. Protein was eluted from the IMAC-column by step elution with 250 mM imidazole, pooled, and dialyzed against Thrombin cleavage buffer (20 mM Tris, 150 mM NaCl, pH 8.0). The His-tag was cleaved off by adding 2 units of thrombin per mg of protein at 4°C overnight. Protein was further purified by an additional IMAC-column and gel filtration (superdex 75) to virtual homogeneity in the presence of reducing agent. Protein was concentrated and stored at −80°C. Identity of the protein was verified by electrospray mass spectrometry. Perdeuterated, isotopically labeled ^{2}H, ^{15}N, ^{13}C-NMR-samples were produced using M9 minimal media made up with ^{2}H$_2$O and ^{13}C-glucose as carbon source and ^{15}NH$_4$Cl as nitrogen source, respectively, and supplemented with vitamin mix.

Crystallization. Protein was rebuffered in 20 mM HEPES, pH 7.4 and concentrated to 50 mg/ml. Crystals were grown by hanging drop vapor diffusion method at 13°C in 24-well crystallization plates. The drops contained 2 μl of protein and 2 μl of reservoir solution (20% glycerol, 2 M ammoniumsulfate, 1% 1,3 butanediol) with 0.5 ml of reservoir solution in each well. Crystals grew within 4 weeks. The structure was determined in house using the anomalous signal of iodine (Single-wavelength Anomalous Dispersion or SAD), after soaking of the crystal in crystallization buffer containing 50 mM KI for 20 h before measurement.

X-Ray Diffraction and Structural Refinement. X-ray diffraction and structural refinement are described in *SI Text*.

CD and Fluorescence. GdmCl ultrapure was purchased from MP Biomedicals, LLC (Eschwege, Germany) and all other chemicals from Merck. All experiments were performed at 15°C in the presence of 0.1 mM TCEP. Far-UV CD GdmCl-induced unfolding transitions of the AR protein were monitored at 222.6 nm for 1–3 μM protein solutions with varying GdmCl concentrations and 4–6 h incubation time to reach equilibrium with a JASCO J600A spectropolarimeter. GdmCl transitions monitored by fluorescence were recorded with a JASCO FP6500 fluorescence spectrometer. A fluorescence spectrum was recorded for each data point from 300 to 420 nm after excitation at 280 nm. All experimental data were analyzed according to a three-state model by nonlinear least-squares fit with proportional weighting to obtain the Gibbs free energy of denaturation ΔG as a function of the GdmCl concentration (31). Fluorescence transition curves detected at various wavelength were analyzed together with CD data using the program Scientist (MicroMath).

Kinetics. Kinetic experiments were performed using an Applied Photophysics SX-17MV and SX-20MV stopped-flow instrument at 15°C. An excitation wavelength of 280 nm was used, and emission was monitored at wavelengths above 305 nm using cut-off filters. Unfolding experiments were performed by mixing native protein or the intermediate state (0 or 1.7 M GdmCl) in 20 mM Na-phosphate (pH 7.4) with 5 or 10 volumes of GdmCl containing the same buffer. Refolding was initiated by 11- or 6-fold dilution of unfolded protein or the intermediate state (5 or 2.6 M GdmCl). The final protein concentration was 1–3 μM. Data collected from at least 4–8 scans were averaged and fitted using Grafit 5 (Erithacus). Unfolding traces were fitted to two exponential functions. Refolding traces were fitted to a sum of three exponential functions. The slowest refolding phase was also determined by manual mixing in the presence and absence of the prolyl isomerase *Thermus thermophilus* SlyD.

NMR. All NMR spectra were acquired with a Bruker Avance 800 and 900 spectrometer in 20 mM Na-phosphate buffer, pH 7.4, containing 10% ^{2}H$_2$O. For backbone assignment trHNCA, trHNCACB and trHN(CO)CACB were measured with a 1.2 mM ^{15}N/^{13}C/^{2}D labeled sample at 25°C. The assignment was transferred by a series of ^{15}N-TROSY-HSQC at different temperatures to 15°C. The GdmCl transition was performed with ^{15}N labeled samples at 15°C using ^{15}N-TROSY-HSQC. Native protein was diluted with 8 M GdmCl stock solution to the desired GdmCl concentration. For each data point in the transition region (1.5–3 M GdmCl), a fresh sample of 500 μM protein was prepared before use to avoid fibril formation (see SI Fig. 11). Signal intensities of all spectra were referenced according to the protein concentration in the respective sample. Spectra were processed using NMRpipe (32) and analyzed with NMRView (33). Signal intensities of the native, intermediate, and unfolded population were used for analysis of the GdmCl equilibrium transition and compared with populations resulting from fluorescence and CD measurements.

Subproject C

ACKNOWLEDGMENTS. We thank Paul Rösch and Hartmut Oschkinat for NMR spectrometer time at 800 and 900 MHz, Franz Xaver Schmid for use of equipment, Peter Schmieder (North-East NMR center, FMP Berlin) for measurement of the assignment spectra, Rolf Sachs and Gerd Hause for electron microscopy, Heinrich Sticht for BlastP searches, and Mirko Sackewitz for helpful discussions. This research was supported by grants from the Deutsche Forschungsgemeinschaft (Ba 1821/3–1,2 and GRK 1026 "Conformational transitions in macromolecular interactions") and the excellence initiative of the state Sachsen-Anhalt.

1. Bork P (1993) *Proteins* 17:363–374.
2. Binz HK, Amstutz P, Kohl A, Stumpp MT, Briand C, Forrer P, Grütter MG, Plückthun A (2004) *Nat Biotechnol* 22:575–582.
3. Mosavi LK, Cammett TJ, Desrosiers DC, Peng ZY (2004) *Protein Sci* 13:1435–1448.
4. Main ER, Lowe AR, Mochrie SG, Jackson SE, Regan L (2005) *Curr Opin Struct Biol* 15:464–471.
5. Gorina S, Pavletich NP (1996) *Science* 274:1001–1005.
6. Howard J, Bechstedt S (2004) *Curr Biol* 14:R224–6.
7. Plaxco KW, Simons KT, Baker D (1998) *J Mol Biol* 277:985–994.
8. Lowe AR, Itzhaki LS (2007) *J Mol Biol* 365:1245–1255.
9. Zweifel ME, Barrick D (2001) *Biochemistry* 40:14357–14367.
10. Devi VS, Binz HK, Stumpp MT, Plückthun A, Bosshard HR, Jelesarov I (2004) *Protein Sci* 13:2864–2870.
11. Tang KS, Guralnick BJ, Wang WK, Fersht AR, Itzhaki LS (1999) *J Mol Biol* 285:1869–1886.
12. Mosavi LK, Williams S, Peng ZY (2002) *J Mol Biol* 320:165–170.
13. Street TO, Bradley CM, Barrick D (2007) *Proc Natl Acad Sci USA* 104:4907–4912.
14. Yuan C, Li J, Selby TL, Byeon IJ, Tsai MD (1999) *J Mol Biol* 294:201–211.
15. Mello CC, Barrick D (2004) *Proc Natl Acad Sci USA* 101:14102–14107.
16. Kloss E, Courtemanche N, Barrick D (2008) *Arch Biochem Biophys*, 469:83–99.
17. Löw C, Weininger U, Zeeb M, Zhang W, Laue ED, Schmid FX, Balbach J (2007) *J Mol Biol* 373:219–231.
18. Zeeb M, Rösner H, Zeslawski W, Canet D, Holak TA, Balbach J (2002) *J Mol Biol* 315:447–457.
19. Mello CC, Bradley CM, Tripp KW, Barrick D (2005) *J Mol Biol* 352:266–281.
20. Kawashima T, Amano N, Koike H, Makino S, Higuchi S, Kawashima-Ohya Y, Watanabe K, Yamazaki M, Kanehori K, Kawamoto T, et al. (2000) *Proc Natl Acad Sci USA* 97:14257–14262.
21. Baumgartner R, Fernandez-Catalan C, Winoto A, Huber R, Engh RA, Holak TA (1998) *Structure* 6:1279–1290.
22. Brotherton DH, Dhanaraj V, Wick S, Brizuela L, Domaille PJ, Volyanik E, Xu X, Parisini E, Smith BO, Archer SJ, et al. (1998) *Nature* 395:244–250.
23. Dobson CM (2003) *Nature* 426:884–890.
24. Fersht AR (2000) *Proc Natl Acad Sci USA* 97:14121–14126.
25. Otzen DE, Kristensen O, Proctor M, Oliveberg M (1999) *Biochemistry* 38:6499–6511.
26. Sanchez IE, Kiefhaber T (2003) *J Mol Biol* 325:367–376.
27. Balbach J, Schmid FX (2000) in *Mechanisms of Protein Folding*, ed Pain RH (Oxford Univ Press, Oxford), pp 212–237.
28. Pappenberger G, Aygun H, Engels JW, Reimer U, Fischer G, Kiefhaber T (2001) *Nat Struct Biol* 8:452–458.
29. Kohl A, Binz HK, Forrer P, Stumpp MT, Plückthun A, Grütter MG (2003) *Proc Natl Acad Sci USA* 100:1700–1705.
30. Mosavi LK, Minor DL, Jr., Peng ZY (2002) *Proc Natl Acad Sci USA* 99:16029–16034.
31. Hecky J, Müller KM (2005) *Biochemistry* 44:12640–12654.
32. Delaglio F, Grzesiek S, Vuister GW, Zhu G, Pfeifer J, Bax A (1995) *J Biomol NMR* 6:277–293.
33. Johnson BA, Blevins RA (1994) *J Biomol NMR* 4:603–614.
34. Koradi R, Billeter M, Wüthrich K (1996) *J Mol Graphics* 14:51–55.

Subproject C

Supporting Information (SI)

Materials and Methods

X-Ray Diffraction and Structural Refinement. A redundant data set from a single iodine derivatized crystal was collected in house at -180 °C with Cu Kα radiation ($\lambda=1.5418$ Å) using a rotating-anode source (RA Micro 007, RigakuMSC) and image plate detector (R-AXIS IV++, RigakuMSC). The crystal diffracted to a resolution of 1.83 Å. Data were indexed and scaled in Mosflm and Scala programs respectively (1-3). The anomalous scattering signal from iodine atoms was used for substructure determination using Single wavelength Anomalous Diffraction (SAD). Eleven iodine positions with occupancy between 1 and 0.05 were determined with the program SHELXD (4) using diffraction data up to 2.3 Å. For phase calculation and further density modification using the program SHELXE, the number of iodine atoms was truncated to 7 and only those with occupancy greater than 0.2 were used. At this stage, the space group could be determined as $P6_422$.

The resulting electron density had excellent quality and was used for automatic main-chain tracing and side-chain docking carried out with the ARP/wARP software (5) and the CCP4 suite (3). The model from the autobuild process comprised two chains (residues from 20 to 188). The missing residues were rebuilt and the structure manually verified against sigma weighted difference Fourier maps using Coot (6) program. Refinement against a maximum likelihood target with a combination of restrained and TLS refinement was performed using Refmac (7). During refinement, 5% of the reflections were randomly chosen and left out for cross-validation using the free R factor. The structure has been refined to an R factor of 18.6% and R_{free} of 20.22% respectively (SI Table I).

A data set from a single non derivatized crystal was first collected in house (resolution 1.75 Å) and later at BESSY to the resolution of 1.65 Å. The synchrotron data were collected at beam line BL 14.1 equipped with a fast scanning 225mm CCD-mosaic detector from MARRESEACH (Norderstedt, Germany) at a wavelength of 0.9123 A. Graphical analysis and refinement was carried out as described for the iodine derivatized crystal. The final model consists of residues 6 to 189, the missing N-terminal amino acids show a non interpretable electron density map and are most likely disordered. There are 497 solvent molecules included in the model, as well as 8 sulphate ions, 3 Cl ions, 4 1,3-butanediol molecules, 6 glycerol molecules and 1 TRIS molecule. For some residues the electron density indicated multiple side chain conformations in the crystal. The model was refined to an R factor of 15.58% and R_{free} of 17.83%. The refined model has good geometry as judged by PROCHECK (3) – Ramachandran statistics showed 90.4 % of amino acids in favoured regions, 9.0 % in allowed regions, 0.6 % generously allowed and none in disallowed regions.

ANS and Thioflavin T Fluorescence. To study ANS binding properties, protein was incubated in 20 mM Na-phosphate, 2 M GdmCl at a concentration of 1 mM at 25 °C for fibril formation to occur. A 1 mM sample without GdmCl was used as a reference. To determine ANS fluorescence, samples were briefly mixed and then diluted to a final concentration of 5 µM with 50 µM ANS in 20 mM Na-phosphate, 2 M GdmCl, pH 7.4. Fluorescence spectra were recorded at an emission wavelength of 410 to 600 nm upon excitation at 370 nm at 25°C. Fibril formation was also monitored by Thioflavin T fluorescence according to standard procedures with minor modifications (8). Optimization of protein concentration, temperature and amount of GdmCl was necessary to avoid fibril formation for later NMR experiments.

Electron Microscopy (EM). For EM analysis, carbonized copper grids (Plano, Wetzlar, Germany) were pretreated for 1 min with bacitracin (0.1 mg/ml). After air drying, protein (preincubated in 2 M GdmCl) that had been diluted with 20 mM Na-phosphate to final concentrations of 0.5 mg/ml was applied for 3 min. Subsequently, grids were again air dried.

Protein (fibrils) was negatively stained with 1% (w/v) uranyl-acetate and visualized in a Zeiss EM 900 electron microscope operating at 80 kV.

Limited proteolysis and Mass Spectrometry. tANK (3 mg/ml) was incubated with 10% Chymotrypsin at 15 °C in 20 mM Na-phosphate. The reaction was stopped at certain time by addition of sample buffer and subsequent boiling. Fragments were separated on a 4-20% gradient SDS gel resulting in 2 distinct fragments with an estimated mass of 18.5 kDa and 10 kDa (data not shown). Stained protein bands were excised, washed and digested with modified trypsin. Peptides were applied to MALDI target plates as described (9). Mass spectra were obtained automatically by MALDI-TOF MS in reflection modes (Voyager-DE-STR), followed by automatic internal calibration using tryptic peptides from autodigestion. The spectra were analysed using the software MoverZ from Genomic Solutions (http://65.219.84.5/moverz.html)) with a signal to noise ratio threshold of 3.0. Resulting peptide mass lists were used to match the thermophilc AR sequence using the Mascot Search engine (www.matrixscience.com).

Ultracentrifugation measurements. Sedimentation analyses were conducted using an analytical ultracentrifuge Optima X-LA (Beckman Instruments, Palo Alto, CA) equipped with two channel cells and an An50 Ti rotor. For these analyses, the protein concentration was varied between 0.1 and 1 mg/ml. Dilutions were performed with 20 mM Na-phosphate, pH 7.4, 1.95 M GdmCl. Native protein was analyzed in the same buffer in the absence of GdmCl. The samples were measured at 20 °C and a wavelength of 280 nm. Sedimentation velocity was carried out at 40,000 rpm, scans were taken every 10 min. Sedimentation equilibrium was performed at 20,000 rpm. The data were analyzed with the software provided by Beckman Instruments. Viscosity and density of the buffer containing 1.95 M GdmCl were calculated as described (10).

SI Figure 7

Fig. 7. Sequence alignment of tANK from *Thermoplasma volcanium* with human CDK inhibitor p19^{INK4d}. Identical and similar residues are colour coded in red. Helices of individual ARs above L36 are indicated.

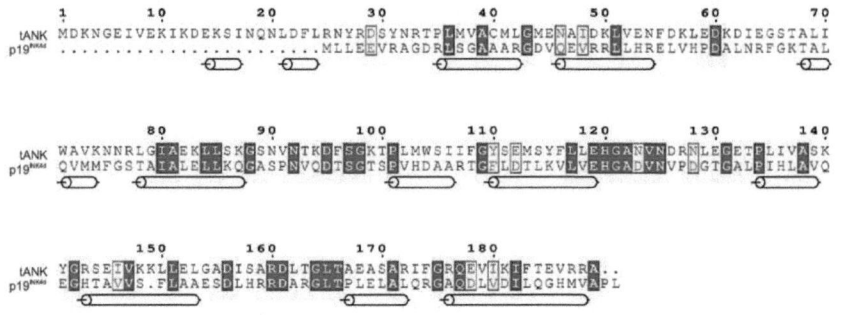

SI Figure 8

Fig. 8. Analytical ultracentrifugation equilibrium sedimentation profile of tANK at 1.95 M GdmCl fits to a single exponential model with $M_r = 22,700 \pm 1,600$, indicating that the intermediate state exists as a monomer in solution. The measurement was performed at 20 °C, 1 mg/ml and 20,000 rpm for 62 h. Residual deviation of the fit from the experimental data is shown at the bottom.

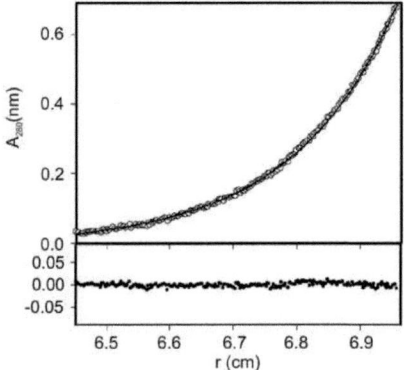

Fig. 9. 2D ^{15}N-TROSY-HSQC spectrum of 2.0 mM ^{2}H/^{15}N tANK at 25°C in 90%/ 10% H$_2$O/^{2}H$_2$O, pH 7.4. The assigned cross-peaks of the amide backbone are labeled using the one-letter amino acid code and the sequence position. Boxes indicate resonance signals which show cross peak intensities below the plotted contour level.

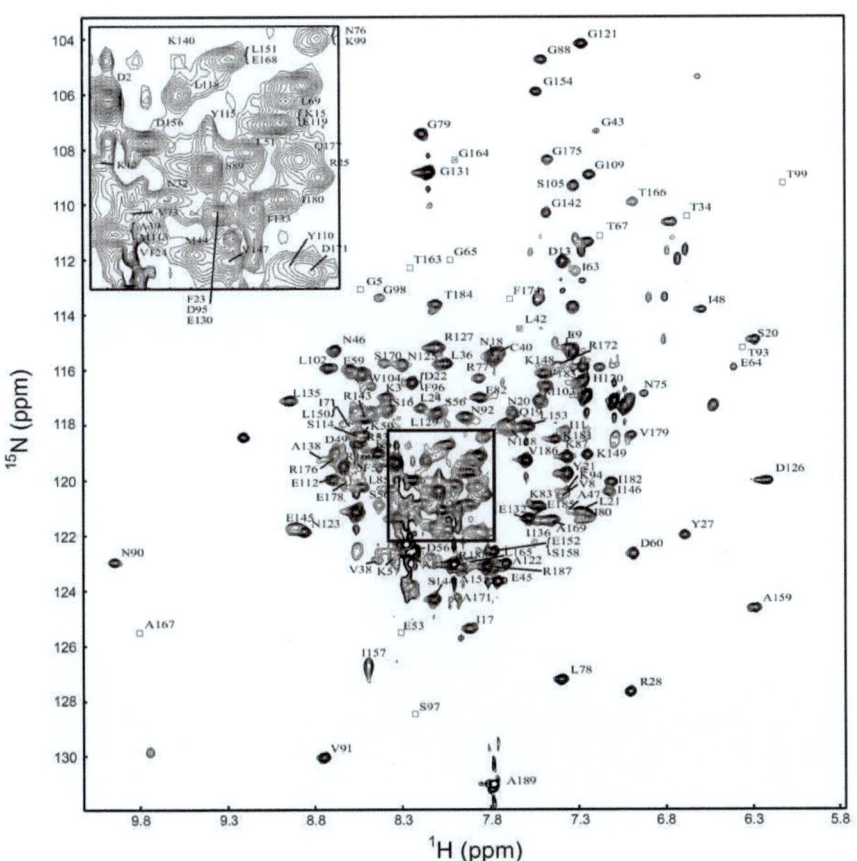

SI Figure 10

Fig. 10. Temperature-dependence of the slowest two apparent rate constants for refolding of tANK. Refolding conditions were 0.9 M GdmCl, pH 7.4. The activation energy (E_A) and the pre-exponential factor A for both reactions were obtained using the Arrhenius equation (solid line). The slowest refolding reaction (open triangles) shows a temperature-independent activation energy of 78.5 ± 3 kJ mol^{-1} ($A = 10^{12.2}$ s^{-1}). The activation parameter for the faster folding reaction (open circles) result in 47.5 ± 3 kJ mol^{-1} ($A = 10^{9.17}$ s^{-1}).

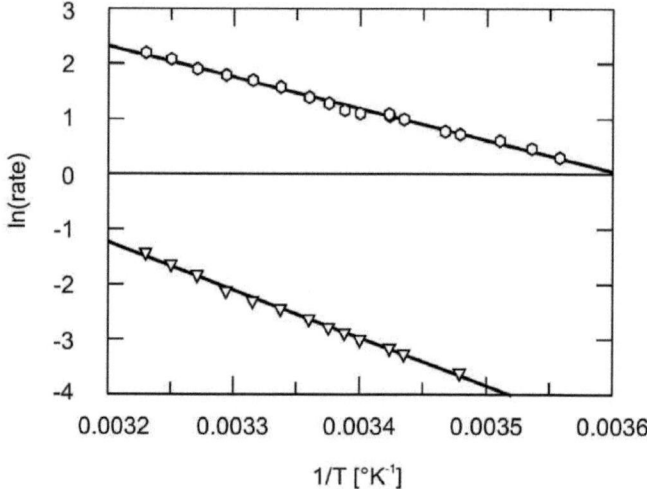

Subproject C

SI Figure 11

Fig. 11. Fibrillation of tANK. 1 mM tANK was incubated at (A) 2 M GdmCl and (B) 0 M GdmCl. Aliquots were transfered at different timepoints to 2 M GdmCl solutions containing ANS, incubated for 5 min and fluorescence spectra were recorded. Preincubated protein at 2 M GdmCl strongly binds ANS (violet line), whereas freshly dissolved protein at 2 M GdmCl does not (black line in (A) and (B)), although the intermediate is maximally populated. (C) Fibrillation recorded by Thioflavin T fluorescence. Kinetics show a lag phase typical for fibril formation. Fibrillation speed strongly depends on concentration (data not shown), temperature and the percentage of populated intermediate. (D) Electron microscopy of negatively stained tANK aggregates. Fibrils were formed by incubation of 1 mM protein in 20 mM Na-phosphate, 2 M GdmCl, pH 7.4 at 25 °C for two hours (scale bar, 200 nm).

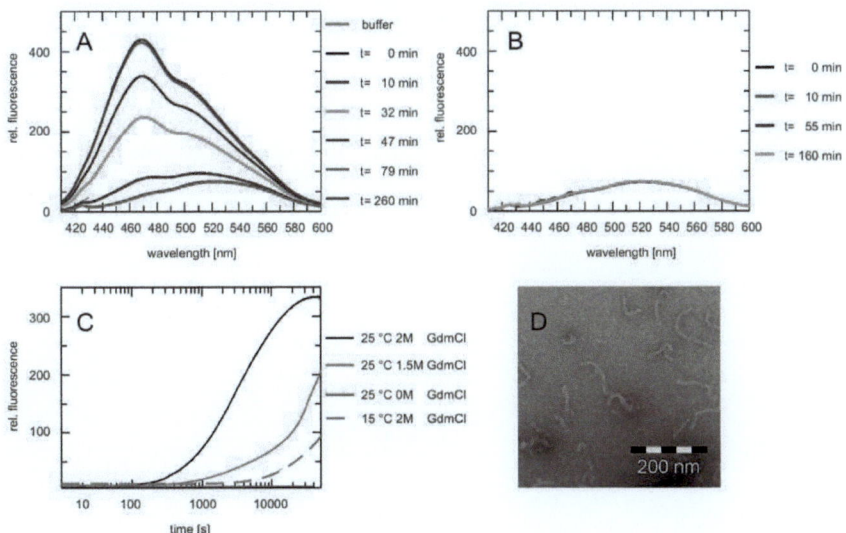

SI Table I

Table I. Data collection and refinement statistics

Data collection	Native	Derivative
Wavelength (Å)	0.9123	1.5418
Space group	P6$_4$22	P6$_4$22
Cell dimensions, Å	a = 100.946 b = 100.946 c = 173.864	a = 101.002 b = 101.002 c = 174.483
Resolution limits, Å	61.66 – 1.65	30.92 – 1.83
No. of reflections	582856	715148
No. of unique reflections	58965	46388
Completeness, %	93.0 (66.9)	98.5 (89.9)
Redundancy	9.9 (6.2)	15.4 (7.2)
<I/σ(I)>	28.1 (5.8)	24.4 (4.7)
Rmerge (%)	0.055 (0.323)	0.095 (0.338)
Rmeas (%)	0.058 (0.353)	0.098 (0.364)
Refinement		
Resolution range (Å)	20.00 –1.65 (1.69-1.65)	
Completeness (working + test) (%)	92.76 (62.83)	
No. of reflections (F>0)	55887 (2737)	
Wilson B (Å^2)	23.5	
Rcryst (%)	15.58 (15.60)	
Rfree (%)	17.83 (20.70)	
No. of non-hydrogen atoms	3725	
Protein	3119	
Water	497	
Sulfate	40	
Glycerol	48	
CL	3	
1,2-butanediol	30	
TRIS	8	
R.m.s.d. from ideality		
Bond lengths (Å)	0.014	
Bond angles (°)	1.373	
Dihedrals (°)	17.3	
Improper (°)	0.33	

Subproject C

Data collection	Native	Derivative
Average B-factor ($Å^2$)	45.82	
Protein atoms	46.99	
Main chain	45.83	
Water	39.71	
glycerol	40.52	
SO_4^{2-}	41.42	
CL	68.70	
1,2-butanediol	39.93	
TRIS	32.22	

$R_{merge} = \Sigma|I - <I>| / \Sigma<I>$.

Values in parentheses correspond to the highest resolution shell for data scaling (1.74−1.65 Å) and refinement (1.69-1.65 Å) respectively. For the derivative dataset, the highest resolution limit is (1.93 – 1.83 Å)

$R_{cryst} = \Sigma||F_{obs}| - |F_{calc}||/\Sigma|F_{obs}|$.

R_{free} is calculated as for R_{cryst} but for a test set comprising reflections not used in the refinement (5.1%).

SI Table 2

Table 2a. Identified peptides of the larger limited proteolysis fragment of tANK (≈ 18.5 kDa)

fragment number	observed mass [dalton]*	calculated mass [dalton]	peptide sequence	sequence position
1	1087.5281	1086.4842	R.NYRDSYNR.T	26-33
2	1402.7586	1401.7504	K.DIEGSTALIWAVK.N	62-74
3	1253.5830	1252.6048	K.GSNVBTKDFSGK.T	88-99
4	1370.7948	1369.7453	R.NLEGETPLIVASK.Y	128-140
5	1746.9964	1745.9312	R.NLEGETPLIVASKYGR.S	128-143
6	951.5327	950.5185	K.YGRSEIVK.K	141-148
7	1285.7499	1284.7401	K.KLLELGADISAR.D	149-160
8	1157.6640	1156.6452	K.LLELGADISAR.D	150-160
9	1089.6127	1088.6342	R.IFGRQEVIK.I	173-181

Table 2b. Identified peptides of the small limited proteolysis fragment of tANK (≈ 10 kDa)

fragment number	observed mass [dalton]*	calculated mass [dalton]	peptide sequence	sequence position
1	1370.8069	1369.7453	R.NLEGETPLIVASK.Y	128-140
2	1747.0270	1745.9312	R.NLEGETPLIVASKYGR.S	128-143
3	1285.8013	1284.7401	K.KLLELGADISAR.D	149-160
4	1157.6884	1156.6452	K.LLELGADISAR.D	150-160
5	2343.3569	2342.2441	K.LLELGADISARDLTGLTAEASAR.I	150-172
6	1204.6607	1203.6095	R.DLTGLTAEASAR.I	161-172
7	1089.6727	1088.6342	R.IFGRQEVIK.I	173-181
8	920.5358	919.5239	K.IFTEVRR.A	182-188

During ionization the peptides get - in case of MALDI - singly protonated. Therefore, the masses get shifted by 1 Da, $[M+H]^+$.

References

1. Leslie, A. G. W. (1992) *Joint CCP4 and ESF-EACMB Newsletter on Protein Crystallography* **26**.
2. Evans, P. R. (1997) *Proceedings of CCP4 Study Weekend on Recent Advances in Phasing*.
3. (1994) *Acta Crystallogr D Biol Crystallogr* **50**, 760-3.
4. Schneider, T. R. & Sheldrick, G. M. (2002) *Acta Crystallogr D Biol Crystallogr* **58**, 1772-9.
5. Perrakis, A., Morris, R. & Lamzin, V. S. (1999) *Nat Struct Biol* **6**, 458-63.
6. Emsley, P. & Cowtan, K. (2004) *Acta Crystallogr D Biol Crystallogr* **60**, 2126-32.
7. Murshudov, G. N., Vagin, A. A. & Dodson, E. J. (1997) *Acta Crystallogr D Biol Crystallogr* **53**, 240-55.
8. LeVine, H., 3rd (1993) *Protein Sci* **2**, 404-10.
9. Peltier, J. B., Emanuelsson, O., Kalume, D. E., Ytterberg, J., Friso, G., Rudella, A., Liberles, D. A., Soderberg, L., Roepstorff, P., von Heijne, G. & van Wijk, K. J. (2002) *Plant Cell* **14**, 211-36.
10. Kawahara, K. & Tanford, C. (1966) *J Biol Chem* **241**, 3228-32.

7.4 Subproject D

D

Subproject D

Subproject D

Crystal Structure and Functional Characterization of the Prolyl Isomerase and Chaperone SlyD from *Thermus thermophilus*

Running title: Crystal structure of thermophilic SlyD

Christian Löw[1], Piotr Neumann[2], Henning Tidow[3], Ulrich Weininger[1], Beatrice Epler[2], Christian Scholz[4], Milton T. Stubbs[2,5], Jochen Balbach[*,1,5]

[1] Institut für Physik, Biophysik, Martin-Luther-Universität Halle-Wittenberg, D-06120 Halle (Saale), Germany

[2] Institut für Biochemie und Biotechnologie, Physikalische Biotechnologie, Martin-Luther-Universität Halle-Wittenberg, D-06120 Halle (Saale), Germany

[3] Medical Research Council Centre for Protein Engineering, Hills Road, Cambridge CB2 0QH, United Kingdom

[4] Roche Diagnostics GmbH, Nonnenwald 2, D-82377 Penzberg, Germany

[5] Mitteldeutsches Zentrum für Struktur und Dynamik der Proteine (MZP), Martin-Luther-Universität Halle-Wittenberg, Germany

*Correspondence should be addressed to

Jochen Balbach	Tel.:	++49 345 55 25353
Institut für Physik, Fachgruppe Biophysik	Fax:	++49 345 55 27383
Martin-Luther-Universität Halle-Wittenberg	e-mail:	jochen.balbach@physik.uni-halle.de
D-06120 Halle(Saale), Germany		

Summary

SlyD is a prolyl isomerase (PPIase) of the FKBP type with chaperone properties. Crystal structures derived from different crystal forms revealed that SlyD from *Thermus thermophilus* (tSlyD) consists of two domains representing the functional units. The PPIase activity is located in a typical FKBP domain, whereas the chaperone function is situated in the autonomiously folded inserted flap domain (IF domain). Both domains display different stabilities according to NMR detected H/D exchange and fluorescence equilibrium transitions. The two isolated domains are stable and functional in solution, but the presence of the IF domain increases the catalytic efficiency of the full length protein towards proline limited refolding of ribonuclease T1 100-fold. Therefore, both domains work synergistically to assist folding of polypeptide chains. The substrate binding surface of tSlyD was mapped by NMR chemical shift perturbations. The combination of folding catalysis with a distal binding site for the folding protein chain is a common principle of this class of enzymes.

Keywords: protein folding; X-ray structure, PPIase-chaperones, FKBP, substrate binding; NMR, SAXS

Abbreviations: SAXS, small-angle X-ray scattering; SlyD, product of the *slyD* (sensitive-to-lysis) gene; SlyD*, 1-165 fragment of *Escherichia coli* SlyD; tSlyD, full length protein of *Thermus thermophilus* SlyD; tSlyDΔIF, variant of tSlyD in which the IF domain is replaced by the flap of human FKBP12; Thermus BP12, variant of FKBP12 in which the flap was replaced by the IF domain of tSlyD; GdmCl, guanidinium chloride; RCM-T1, reduced and carboxymethylated form of the S54G/P55N double mutant of ribonuclease T_1; RCM-α-lactalbumin, reduced and carboxymethylated form of α-lactalbumin; Tat, twin-arginine translocation.

Introduction

Protein folding reactions of single domain proteins can be in some instances very fast (within milliseconds) and effective [1-4]. However, folding of many proteins is much slower than expected, because it is coupled to slow steps such as prolyl isomerization [5, 6] or disulfide bond formation [7, 8]. During slow folding reactions, intermediates states can accumulate raising the risk of misfolding and aggregation [9-12]. To suppress this process and to speed up productive protein folding, nature has evolved folding helper proteins. Disulfide oxido reductases and peptidyl-prolyl *cis/trans* isomerases (PPIases) catalyze slow protein folding reactions and thus reduce the lifetime of aggregation-prone folding intermediates [13, 14]. Chaperones bind hydrophobic patches of unfolded or partially folded polypeptide chains and prevent their aggregation [15-18]. Several folding helper proteins combine both of these functions. FkpA [19-21], Trigger factor [22-24], SurA [25], and SlyD [26, 27] show prolyl isomerase and chaperone activity. The combination of folding activity with a separate binding site for the folding protein chain is also observed for disulfide isomerases (PDIs) [28, 29].

The enzymatic and chaperone-like acitivities are typically localised in different domains, but only little is known about their interplay. SlyD shows this dual topology and belongs to the group of PPIases [27, 30]. Three families of this enzyme group are known: cyclophilins, FK506 binding proteins (FKBPs), and parvulins [31, 32]. Despite their low sequence and structure similarity, PPIases can catalyze the isomerization of prolyl peptide bonds and some are involved in signal transduction, protein assembly, or cell cycle regulation [32-34].

For the FKBP familiy, binding of the immunosuppressant FK506 is characteristic [35]. Their members are widely distributed among all kingdoms of life and occur as single domain proteins or as units in larger multidomain proteins [33, 36]. Human FKBP12 shows a comparably low catalytic activity [37], which could be increased 200-fold by the addition of the chaperone domain of *Escherichia coli* SlyD [38], which provides additional binding sites for folding polypeptide chains.

The cytoplasmic prolyl *cis/trans* isomerase SlyD was discovered in *E. coli* as a host factor for the ΦX174 lysis cycle [39-41]. It pressumably stabilizes the viral lysis protein E. Certain mutations in the *slyd* gene lead to *E. coli* strains, which are resistant to lysis by the bacteriophage giving raise to its acronyme 'sensitive to lysis D' [42]. Little is known about the detailed physiological function of SlyD [43-45]. Recent investigations suggest a participation of SlyD in the Tat (Twin arginine translocation) transport system as binding partner of the hydrophobic Twin arginine signal peptide [46]. Earlier findings, that SlyD might act as a metallochaperone in the hydrogenase biosynthetic pathway [30], were recently confirmed [45]. Proteome analysis of *E. coli* BL21(DE3) after heat shock [47] or membrane protein overexpression [48] revealed significant increase of SlyD compared to non-stress conditions, indicating that SlyD facilitates folding and might increase the solubility of many aggregation prone proteins in the *E. coli* cytoplasma. Covalent fusion of aggregation prone proteins with SlyD modules showed enhanced cytosolic expression and solubility [26, 49].

SlyD has a 45 amino acid insertion in the flap region close to the prolyl isomerase active site, which is missing in human FKBP12. Here we present the crystal structure and biophysical characterization of SlyD from *Thermus thermophilus*, which shows more than 50 percent sequence identity to *E. coli* SlyD* (1-165). The structure of tSlyD comprises a FKBP like domain and the insertion in the flap region folds into an autonomous domain (IF domain), similar to MtFKBP17 [50]. The protein crystallized under two conditions in two different space groups and the corresponding structures are altered regarding the relative orientation of the IF and FKBP domain. This observation indicates that both domains are flexible and independent towards each other in solution. NMR H/D exchange spectroscopy revealed a significant higher stability for the FKBP domain compared to the IF domain. Both domains are stable and functional in isolation and do not interact with each other. However, the presence of the chaperone domain increases in wild type tSlyD the isomerase activity 100-fold for protein substrates. Using NMR chemical shift analysis with various unfolded substrates, we identified the binding region for unfolded polypeptide chains in the IF domain. Binding is driven by hydrophobic interactions. The structure and biophysical data of tSlyD propose that both domains work synergistically to facilitate folding of polypeptides.

Results

X-ray structure and SAXS analysis of *Thermus thermophilus* SlyD (tSlyD)

Sequence alignment of tSlyD and *E. coli* SlyD* (1-165) revealed homology with more than 50 % sequence identity (Fig. 1a). Full length *E. coli* SlyD (1-196) carries an additional cysteine- and histidine-rich C-terminal tail, which is proposed to act as metal binding site, but is absent in tSlyD. The deletion of this C-terminal part was shown to have no influence on the prolyl isomerase and chaperone activity in *E. coli* SlyD* [27]. The three dimensional structure of tSlyD was determined by X-ray crystallography and structural statistics are summarized in Table 1. tSlyD is a two domain protein consisting of a FKBP domain, typical for PPIases, and an autonomiously folded IF domain (Fig. 1b-d). Crystals were obtained under two different conditions and the resulting structures show variability regarding the domain orientation. The overall structure of the domains is unchanged in both structures except for ß8 and ß9 of the IF domain being shorter in the crystal structure obtained from crystal form B (Fig. 1c). However, in the crystal structure derived from crystal form A (Fig. 1b), the domains are in closer proximity, caused by a twist and movement of the linker region (Fig. 1d). Crystal packing for both structures is significantly different. Characteristic for crystal form A is a high solvent content of about 72 %. tSlyd molecules form here rings lying one above each other when viewed along crystallographic z-axis. These rings exhibit a huge solvent channel of a diameter of about 90 Å. The solvent accessible surface of such rings is mostly formed by the C-terminus, the internal surface of the C-shape like IF domain and a loop comprising residues 79 to 85 of the FKBP domain (data not shown).

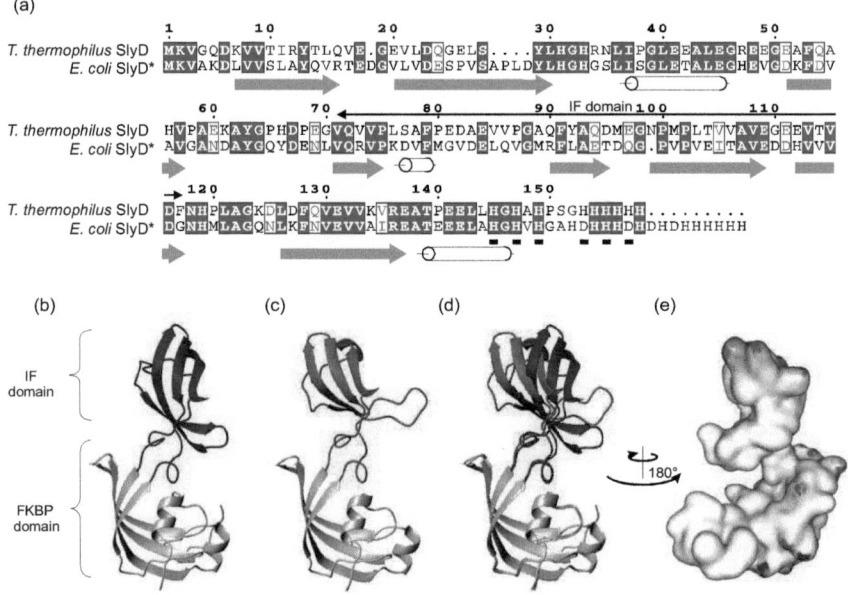

Figure 1. (a) Sequence alignment of tSlyD and *E. coli* SlyD*. Identical and similar residues between these two proteins are boxed or coloured in red, respectively. Both constructs carry a C-terminal histidine-tag. Histidine residues involved in Ni^{2+} ion binding are underlined. Secondary structure elements and localisation of the IF domain are indicated. (b,c) Schematic representation of the crystal structure of tSlyD (3CGM.pdb, 3CGN.pdb in Protein Data Base) derived from two different crystal forms (crystal form A (b), crystal form B (c)). The FKBP and the IF domain are indicated. (d) Superposition of the crystal structures of tSlyD demonstrates the different orientation of the two domains towards each other. The C_α atoms of the FKBP domain were used for superimposition. (e) Electrostatic surface potential representation of tSlyD (crystal form A). Negative potentials are shown in red, positive potentials in blue, and neutral in white. The figures were created using MOLMOL [51].

A detailed crystal packing analysis of crystal form A revealed that the IF domain forms two hydrophobic and two weak polar interactions with only one symmetry related IF domain (middle part of the ß-strands: residues Gly88 to Asp116, Gln90 to Thr104 and Pro120) and 17 interactions (five of them are polar interactions) mostly with loop regions of the two surrounding FKBP domains. Further contacts of the IF domain to neigbouring molecules are mainly mediated by the outer surface of the C-shape like IF domain, whereas the hydrophobic cleft formed by the internal surface of the domain remains solvent exposed. Significant difference in the average atomic B-factor of both domains (35.2 Å^2 for the FKBP domain and 53.9 Å^2 for the IF domain) indicate higher mobility for the IF domain. The small number of contacts in the crystal lattice and the large solvent exposed surface of the IF domain (IF domain burries 22 % (846 Å^2) of the total (3802 Å^2) solvent accessible area as calculated with Arealmol support this hypothesis. Therefore we conclude that the orientation of the IF domain relative to the FKBP domain is not strongly influenced by crystal packing.

In contrast, the crystal of crystal form B is much more densely packed in the crystal lattice with a solvent content of only 45 %. The IF domain forms 34 interactions with neighbouring IF domains (eight of them are hydrogen bonds and 4 polar interactions) and just four weak interactions with surrounding FKBP domains (one polar and three hydrophobic interactions). The IF domain burries 35 % (1398 Å^2) of the total (4016 Å^2) solvent accessible surface to form crystal contacts mostly with one IF domain positioned in a face to face mode by the crystallographic two-fold axis. In contrast to crystal form A, the two symmetry related IF domains of crystal form B interact with each other using the internal surface of the C-shape like IF domain. The loop comprising residues 93-99 of one IF domain is positioned between the equivalent loop and a short helix (residues 75-79) of the symmetry related IF domain and vice versa. In particular, the side chain of Met96 penetrates into the hydrophobic cleft of the symmetry related IF domain. Hence, the relative orientation of the two domains is strongly influenced by the crystal symmetry.

The IF domain consists of a four-stranded β-sheet and a short α-helix (Fig. 2a) with a predominantly hydrophobic surface, indicating that this part of the molecule is able to bind to hydrophobic patches of unfolded substrates to prevent their aggregation. This assumption is confirmed by a deletion variant tSlyDΔIF lacking the IF domain and thus loosing the chaperone activity (SI Fig. 11). In both structures, the loop region (residues 96-98) in the IF domain which connects β-strand β8 with β-strand β9 has been refined with very high B-factors, implicating high mobility. This is typical for highly flexible loops and agrees well with dynamic studies by NMR (M. Kovermann and J.Balbach, unpublished results).

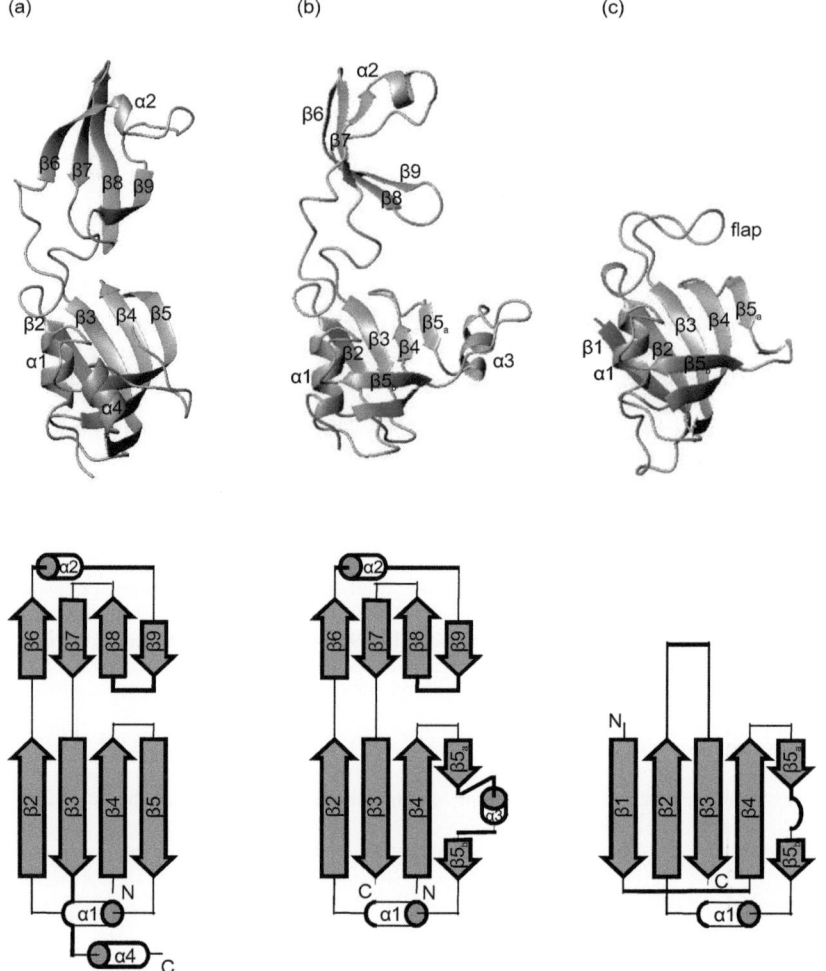

Figure 2. (a) Topological and schematic comparison of tSlyD (a), MtFKBP17 (pdb.1IX5) (b) and HsFKBP12 (pdb.1FKF) (c). Corresponding secondary structure elements in the FKBP and IF domain are labeled.

The PPIase domain of tSlyD shows a characterisitc FKBP-like fold consisting of a four-stranded antiparallel β-sheet and two α-helices. In comparison to human FKBP12 it lacks the N-terminal β-strand (β1) as expected from the sequence alignment. Besides, β-strand β5 is not interrupted by a loop or a helical insertion as found in the archaeon homolog MtFKBP17 and FKBP12 (Fig. 2). Therefore, the substrate binding pocket of tSlyD is smaller. (e.g. R42 is missing in tSylD), which might be the reason for the reduced prolyl isomerase activity of the isolated FKBP domain (Table 2). An additional α-helix at the C-terminus (α4) with unknown function is unique for tSlyD and also expected for *E. coli* SlyD* (NMR structure published elsewhere).

Although tSlyD is devoid of the putative metal binding site proposed for the cysteine- and histidine-rich C-terminal part in *E. coli* SlyD, it is able to bind divalent ions like Ni^{2+} or Zn^{2+} [30]. The tSlyD structure (crystal form A) exhibits a sixfold coordinated Ni^{2+} ion close to the C-terminus (SI Fig. 8), absent in crystal form B. The Ni^{2+} ion is coordinated by three histidines of the histidine-tag and 3 histidines of the histidine-rich motif of the tSlyD sequence. Interestingly, this C-terminal motif is higly conserved among the SlyD family.

Since crystal structures display a static picture of highly dynamic proteins, the question raises whether SlyD shows a preferred domain orientation in solution or allows an equal distribution of domain arrangements. Small-angle X-ray scattering (SAXS) was used to characterize the structural features of tSlyD in solution. SAXS experiments were performed on monodisperse solutions of free tSlyD and the tSlyD/Tat-signalpeptide complex (SI Fig. 10). The processed experimental patterns (scattering intensity *I versus* momentum transfer $s = 4\pi \sin\theta/\lambda$, where 2θ corresponds to the scattering angle and $\lambda = 1.5$ Å to the wavelength) are displayed in Fig. 3.

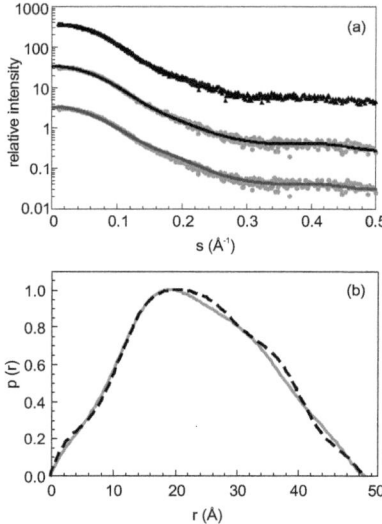

Figure 3. SAXS analysis of tSlyD. (a) Experimental intensities for free tSlyD (light grey) and a tSlyD/Tat-signalpeptide complex (black triangles). The scattering profiles are displaced along the ordinate for better visualization. The fit to the scattering pattern computed from the crystal structures of tSlyD is shown in black (chi = 1.61) (crystal form A, middle) and grey (chi= 2.43) (crystal form B, bottom). (b) Distance distribution plots computed from the experimental data and normalized to the maximum value of unity. Colour code as in (a).

In general, the experimental intensities and fits computed from both tSlyD crystal structures agree well with the experimental SAXS data in solution. However, the structure derived from crystal form A shows a significantly better fit than the one derived from crystal form B (chi = 1.6 vs. chi = 2.4) (Fig. 3a). Thus, the structure from crystal form A seems to better represent the structure in solution. As SAXS data always represent an average over a conformational ensemble, it cannot be ruled out that tSlyD exists in multiple conformations. A low-resolution model of tSlyD generated *ab initio* by the program DAMMIN [52] generated a shape that is consistent with the crystal structure (SI Fig. 10c).

The presence of the Tat-signalpeptide in the complex with tSlyD does not cause a major change in the overall shape of tSlyD according to the minor changes in scattering pattern, distance distribution function and Kratky-plot (Fig. 3 and SI Fig. 10).

Domains of tSlyD display different stabilities

The variability of the domain orientations in the crystals of tSlyD indicates some degree of flexibility of both domains towards each other in solution. To investigate the domain interplay thermodynamically, we cloned, expressed, and purified the domains seperately (named tSlyDΔIF and tIF). CD and NMR spectra confirmed that these constructs were folded (data not shown). Both domains of tSlyD are devoid of tryptophan residues, but contain three tyrosine residues in the FKBP and one in the IF domain. The stabilities of tSlyD and isolated domains were determined by fluorescence spectroscopy. Tyr-fluorescence increased upon unfolding for full length tSlyD and tSlyDΔIF, but was quenched upon unfolding for the isolated tIF domain (Fig. 4).

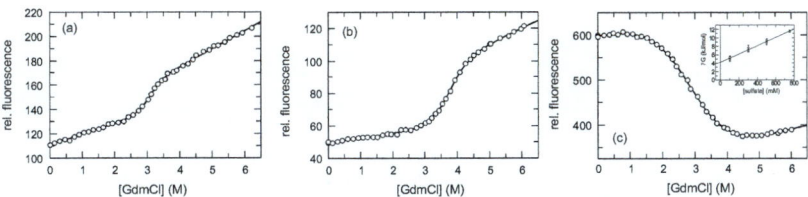

Figure 4. GdmCl induced unfolding transitions of tSlyD (a), tSlyDΔIF (b) and tIF (c). Transition curves of tIF were measured in the presence of various amounts of sulfate (shown transition curve was recorded in the presence of 800 mM sulfate). Resulting $\Delta G_D(H_2O)$ values at each sulfate concentration are shown as inset and the $\Delta G_D(H_2O)$ value is linear extrapolated to zero molar sulfate concentration (c). Unfolding was measured at 25 °C in 50 mM sodium phosphate buffer, 100 mM NaCl, pH 7.5 by the change in fluorescence at 308 nm after excitation at 276 nm. The thermodynamic parameters from the two-state analysis of the transitions are given in Table 2.

Unfolding curves followed an apparent two-state transition and were analyzed according to a two-state model with the resulting biophysical parameters given in Table 2. $\Delta G_D(H_2O)$ values for tSlyD with and without the IF domain were very similar, but due to the decreased m-value for tSlyDΔIF its unfolding midpoint is 0.6 M GdmCl increased. Compared to the mesophilic *E. coli* SlyD [38], the stability of tSlyD is twice as high. The isolated tIF domain was significantly less stable compared to the FKBP domain with a much smaller m-value, as expected for the 52 residues comprising polypeptide. Because of the low thermodynamic stability, GdmCl induced transition curves of tIF were recorded in the presence of various amounts of sulfate and the stability was extrapolated to 0 M sulfate (see inset Fig. 4c). The tIF domain showed a comparatively low stability of 4.3 ± 0.3 kJ/mol in the absence of salt, resulting in a population of 15 % unfolded molecules under native conditions. The 2D $^1H/^{15}N$-HSQC spectrum of ^{15}N labeled tIF domain aquired in the absence of salt showed two populations of peaks, corresponding to the native and the unfolded state of the protein. Unfolded cross-peaks vanished upon addition of salt (data not shown). Despite the low stability, the isolated tIF domain was functional and repressed aggregation (SI Fig. 11) according to a insulin aggregation assay [53].

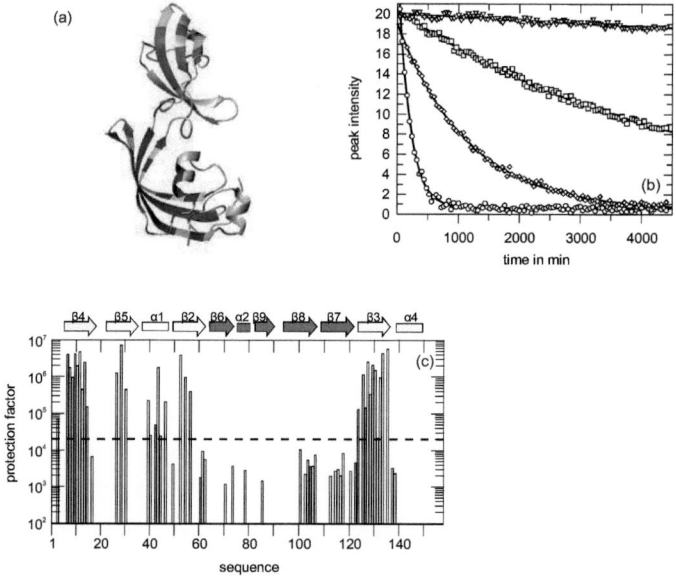

Figure 5. Local stability of tSlyD measured by H/D exchange (a) Residues with highly protected amide protons against solvent exchange (P > 20,000) are indicated in blue and less protected backbone amides (P < 20,000) in red. (b) Examples of H/D exchange curves for individual residues (○ V74, ◊ V132, □ A55, ▽ F53). (c) Protection factors of backbone amides of tSlyD determined by H/D exchange monitored by NMR. Missing bars indicate fast-exchanging amides, amides with missing assignments, and proline residues. Grey symbols for the secondary elements indicate the IF domain.

More detailed information about the stability of the two domains in the full length protein results from an NMR H/D-exchange experiment, which is a sensitive tool to measure the local stability (Fig. 5). NMR resonances of backbone amide protons of the native state were assigned using standard 3D experiments (SI Fig. 9). Amides of the FKBP domain were, on average, 1000 fold more protected than those of the IF domain. This indicates that the latter undergoes frequent local opening without unfolding the FKBP domain and is significantly less stable than the FKBP domain. The four stranded β-sheet of the FKBP domain showed the highest protection factors and these amides exchanged only after global unfolding of the entire protein. Resulting $\Delta G_D(H_2O)$ values of the most protected amides in the FKBP domain (32-37 kJ/mol) correspond to $\Delta G_D(H_2O)$ of tSlyD as derived from the GdmCl induced unfolding transition. Amides of the IF domain, however, were less well protected and resulted in a stability of 20-22 kJ/mol under native conditions.

Activity of tSlyD

tSlyD is a catalyst in proline-limited folding reactions. We used the reduced and carboxymethylated form of the S54G/P55N variant of ribonuclease T1 (RCM-T1) as a model substrate, because its refolding under high-salt conditions is limited by a slow *trans-cis* isomerzation reaction of Pro39. The rate constants of the catalyzed folding reactions depend linearly on the concentrations of the prolyl isomerase (SI Fig. 12). The catalytic efficiencies (k_{cat}/K_M) are given in Table 2. Compared to *E. coli* SlyD* ($k_{cat}/K_M = 0.82 \cdot 10^6$ M^{-1}s^{-1}) [38] the activity of tSlyd was 2.8-fold lower at 15 °C. The k_{cat}/K_M-value of tSlyDΔIF was 100-fold lower compared to the full length protein and 6-fold lower compared to human FKBP12. The tIF domain did not catalyze the refolding of RCM-T1 at all. The insertion of the tIF domain into human FKBP12, however, led to a chimeric protein (named Thermus BP12) with a k_{cat}/K_M-value 180-fold higher than FKBP12 and 10-fold higher than tSlyD. It should be noted that both Thermus BP12 and FKBP12 carried a C22A mutation for this comparison to facilitate long term stability. This strategy was already successfully applied for the *E. coli* IF domain with similar results [38].

Interactions of tSlyD with partially folded and unfolded polypeptides

For a detailed analysis of binding of unfolded or partially folded substrates to tSlyD various substrates were tested. First, RCM-α-lactalbumin was choosen, because the reduced and carboxymethylated form of α-lactalbumin is only partially folded and remains soluble at high concentrations [54]. Second, we used a 27 residue long Tat-signalpeptide (twin-arginine translocation) with the characteristic twin arginine motive, because recent work predicted a role for SlyD in the Tat dependent translocation pathway [46]. Complex formation was monitored by fluorescence and NMR-spectroscopy. Trp-fluorescence of RCM-α-lactalbumin as well as the Tat-signalpeptide increased upon binding of tSlyD, presumably due to shielding of the fluorophores from the solvent (Fig. 6). Binding isotherms resulted in K_D values of 0.5 µM for RCM-α-lactalbumin and about 0.1 µM for the Tat-signalpeptide, respectively. The latter showed a 1:1 binding stoichiometry. tSlydΔIF did not bind to any of these substrates and the affinities for the tIF domain were slightly weaker (Table 2). Major rearrangements of the secondary structure of tSlyD or the substrate upon complex formation were excluded by far UV-CD spectroscopy and SAXS analysis (Fig. 3). The CD spectra of the tSlyD complexes were a linear combination of the spectra of free tSlyD and free substrates (data not shown).

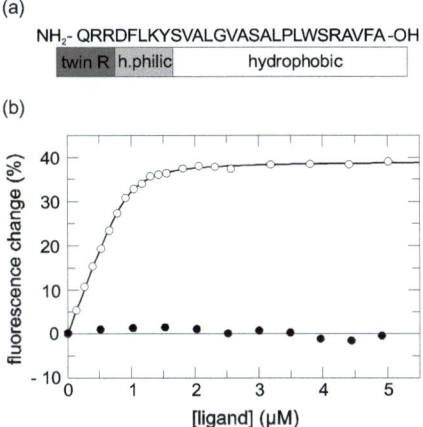

Figure 6 (a) Sequence and schematic representation of the Tat-signalpeptide (b) Fluorescence increase of Trp 21 of the Tat-signalpeptide upon binding to tSlyD (○). tSlyDΔIF (●) did not bind. Binding affinities were determined in 50 mM sodium phosphate, 100 mM NaCl, pH 7.5 at 25 °C. The resulting K_D values are given in Table 3.

High resolution information on the substrate binding site of tSlyD was obtained by 2D $^1H/^{15}N$-HSQC NMR-spectroscopy. NMR titration experiments in the presence of higher concentrations of RCM-α-lactalbumin reveal resonance broadening due to the increased molecular mass of the complex (Fig. 7a). Furthermore, this stoichiometric titration probed by NMR revealed that two tSlyD molecules bind to one RCM-α-lactalbumin polypeptidechain, resulting in a 46 kDa complex (Fig. 7c). Nevertheless, a large number of resonances could be followed over the entire titration range and the mean weighted difference in chemical shifts Δδ for each residue was plotted against the sequence (Fig. 7d). Predominantly hydrophobic residues of the IF domain are involved in the interaction with unfolded polypeptide chains (G70, S77, A78, E81, A83, V85, V86, A89, M96, V106). The binding interface of tSlyD with RCM-α-lactalbumin is visualized in Fig. 7b. Interestingly, D23 of the FKBP domain, which is conserved in FK506 binding proteins, also showed a significant chemical shift change upon substrate binding.

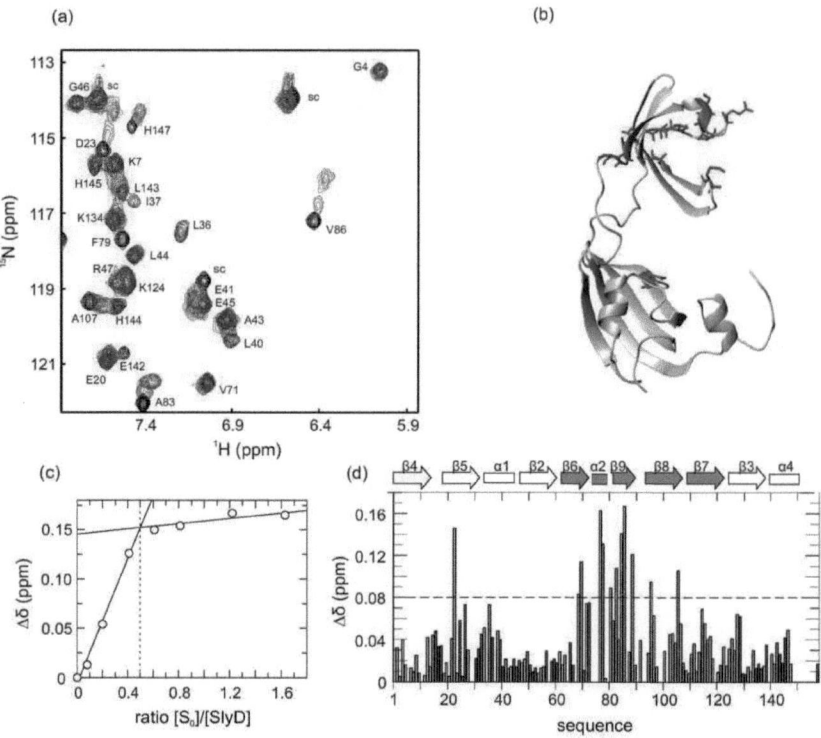

Figure 7 (a) Section of 2D ^{1}H/^{15}N-HSQC spectrum of ^{15}N-tSlyD in the absence (black) and increasing concentrations of unlabeled RCM-α-lactalbumin (from blue to green and red; (1: 1.6 final ratio) at pH 7.5 and 25 °C. The assigned cross-peaks of the amide backbone are labeled using the one-letter amino acid code and the sequence position (cross-peaks of sidechain protons are labeled with sc). Alterations of the resonance frequency are small (G4), medium (H147), or large (V86). Some of the cross-peaks in the complexed protein are broadened beyond detection (F79). (b) Residues of tSlyD showing chemical shift changes Δδ > 0.08 ppm upon binding to RCM-α-lactalbumin are colored in red. (c) Mean weighted difference in chemical shifts (Δδ) for V86 implies that two tSlyD molecules bind to one RCM-α-lactalbumin polypeptide chain. (d) Backbone chemical shift changes (Δδ) of tSlyD upon binding to RCM-α-lactalbumin. Grey symbols for the secondary elements indicate the IF domain.

Discussion and Conclusion

SlyD is a prolyl isomerase of the FKBP type with chaperone properties. The crystal structure revealed that SlyD of *Thermus thermophilus* consists of two different domains, each hosting one functional unit. The FKBP domain of tSlyD is similar to human FKBP12, but lacks the first β-strand, and β-strand 5 is not interrupted by a loop. In addition, tSlyD has an insertion in the flap region, which folds into the seperate IF domain, related to the archaeon homolog MtFKBP12. Using chemical shift mapping, we identified a hydrophobic binding interface for unfolded or partially folded polypeptide chains in the IF domain. The chaperone-like activity is directly related to the IF domain, because the deletion mutant tSlyDΔIF only looses its chaperone activity while keeping its PPIase activity. The presence of the IF domain increases the PPIase activity 100-fold. Hence, the IF domain provides a binding site for folding polypeptides. The combination of a catalytic active domain with a protein binding or chaperone domain is widely used in nature [21, 24, 25, 28]. But the communication of these domains is not well understood. Although the isolated domains of tSlyD are stable and functional, they do not interact with each other. However, the protein crystallized in two forms and the resulting structures are different in terms of domain orientation. The hydrophobic binding interface of the IF domain is more distant from the binding pocket of the FKBP domain in the structure of crystal form B compared to A.

The structure of homologous MtFKBP17 was solved by NMR spectroscopy [50]. The ten lowest energy structures show a poor definition between the two domains, suggesting either a flexible linker between these domains, or insufficient experimental restraints. In summary, we propose a 'swinging-arm' like mechanism. Both domains are flexible and dynamic relative to each other in solution. The IF domain acts as binding site for folding polypeptide chains and aligns them close to the catalytic site. After isomerization, the substrate is released and folds to completion.

Materials and Methods

Materials and reagents

GdmCl ultrapure was purchased from MP Biomedicals, LLC (Eschwege, Germany). All other chemicals were of analytical grade and from Merck (Darmstadt, Germany) or Sigma Aldrich (St. Louis, MO). Tat-signalpeptides were from Activotec (Cambrdige, UK) and α-lactalbumin from Sigma Aldrich (St. Louis, MO). (S54G,P55N)-RCM-RNase T1 was purified as previously described [55]. The permanently unfolded proteins (S54G,P55N)-RCM-RNase T1 and RCM-α-lactalbumin were prepared by reduction and carboxymethylation as described for wild type RNase T1 [55].

Gene construction, protein expression, and purification

The sequence of SlyD from *Thermus thermophilus* (accession number **Q5SLE7**) was retrieved from the SwissProt database. A synthetic gene, encoding tSlyD was purchased from Medigenomix (Martinsried, Germany) and cloned into a pET11a expression vector (Novagen, Madison, WI). The codon usage was optimized for expression in *E. coli* host cells. tSlyDΔIF was constructed by replacing the IF domain with the shorter flap sequence of human FKBP12 [38]. Thermus BP12 was designed according to a described construct (FKBP12 + IF) [38], but using the *Thermus thermophilus* IF domain. The isolated tIF domain covered the sequence from position 65-117 of tSlyD. All recombinant tSlyD constructs contained a C-terminal hexahistidine tag (with an additional small linker consisting of the amino acids PSG before the tag) to facilitate IMAC-assisted purification and refolding. tSlyD variants were produced and purified as previously described with minor modifications [27]. Identity of the protein was verified by electrospray mass spectrometry. Isotopically labeled ^{15}N, ^{13}C-NMR samples were produced using M9 minimal media made up with ^{13}C-glucose as carbon source and ^{15}NH$_4$Cl as nitrogen source, respectively, and supplemented with vitamin mix.

Crystallization and structure determination

Protein was rebuffered in 20 mM HEPES, pH 7.4 and concentrated to 20 - 60 mg/ml. Crystals were grown by hanging drop vapor diffusion method at 13 °C in 24-well crystallization plates. The drops contained 2 µl of protein and 2 µl of reservoir solution (crystal form A: 2 M ammoniumsulfate, 2 % PEG 400, 1 % thiocyanate, 100 mM HEPES, pH 7.5; crystal form B: 1.6 M Na/K tatrate, 100 mM MES, pH 6.5) with 0.5 ml of reservoir solution in each well. Crystals grew within 3 month. The structure of *Thermus thermophilus* SlyD (crystal form A) has been determined by Molecular Replacement (MR) method with data between 30 and 2.5 Å. The FKBP domain of FKBP12 (1D6O.pdb) has been used as the search model with the PHASER program. The MR solution clearly showed that the asymmetric unit comprises only one tSlyD molecule (corresponding solvent contents of 72.7%). The structure has been manually rebuild (missing IF domain) against sigma a weighted difference Fourier maps and verified against simulated annealing (SA) omit maps using O and Coot programs [56]. The structure was refined in CNS program against maximum likelihood target using structure factors (mlf). During refinement 7.5% of the reflections were randomly chosen and left out for cross-validation using the free R factor. The refinement was based on slow cooling SA (both Torsion Angle Dynamics and Cartesian Dynamics) combined with standard minimization and individually restrained B-factor refinement. Both, overall anisotropic B-factor and bulk solvent corrections were applied. The final refinement using combination of restrained and TLS refinement was performed in Refmac [57]. Both programs used the same subset of data for Rfree calculation. The tSlyD structure (crystal form A) has been refined to the R and Rfree of 0.1870 and 0.2257, respectively.

The structure of the second crystal form (crystal form B) has also been determined by MR. Separate FKBP and IF domain of *Thermus thermophilus* SlyD (crystal form A; 3CGM.pdb) have been used as the search models with the PHASER program. The asymmetric unit comprises one tSlyD molecule. The structure of tSlyD (crystal form B) has been refined in similar way as described for crystal form A. The final R and Rfree are 0.210 and 0.256, respectively.

The refined crystal structures of tSlyD have good geometry as judged by PROCHECK program [58] – Ramachandran statistics showed 91.4 % of amino acids in favoured regions, 7.8 % in allowed regions, 0.8 % generously allowed regions for tSlyD derived from crystal form A and 82.1 % of amino acids in favoured regions, 15.4 % in allowed regions, 2.4 % generously allowed regions for tSlyD derived from crystal form B.

All data sets have been collected at BESSY. The used wavelength was 0.91480 Å. The data for tSlyD (crystal form A) were collected at beam line BL 14.1 equipped with a fast scanning 225mm CCD-mosaic detector from MARRESEACH. The second tSlyD crystal (crystal form B) was measured at the beam line BL 14.2 equipped with MARCCD-165 mm detector. The data sets of both crystals were processed, reduced and scaled using the XDS package.

Small-angle X-ray scattering (SAXS)

SAXS data were collected at the X33 beamline at EMBL/DESY, Hamburg following standard procedures [59]. Repetitive data collection on the same sample was performed to monitor possible radiation damage, and no damage was detected. The camera length was 2.7 m at X33 covering the range of scattering vectors 0.012 < s < 0.47 Å^{-1} at a wavelength λ = 1.5 Å. The samples were prepared in concentration ranges from 6.8 to 15 mg/ml in 50 mM sodium phosphate (pH 7.5), 100 mM NaCl. The data was processed using PRIMUS [60]. The radius of gyration (R_g) was evaluated using the Guinier approximation [61] ($I(s) = I(0)\exp(-s^2 R_g^2 / 3)$ for sR_g < 1.3) and from the entire scattering curve with the program GNOM [62]. The latter also provided distance distribution function $p(r)$ and the maximum dimension D_{max}. The M_r of the solutes were evaluated by comparison of the forward scattering intensity with that from a BSA reference solution (M_r = 66 kDa).

Ab initio modelling of SAXS data

Low-resolution models were obtained using the *ab initio* simulated annealing (SA) program DAMMIN [52]. The program generates models consisting of dummy residues or beads, respectively, to fit the experimental data $I_{exp}(s)$ by minimizing the discrepancy:

$$\chi^2 = \frac{1}{N-1} \sum_j \left[\frac{I_{exp}(s_j) - cI_{calc}(s_j)}{\sigma(s_j)} \right]^2 \quad (1)$$

Here N is the number of experimental points, c is a scaling factor and $I_{calc}(s_j)$ and $\sigma(s_j)$ are the calculated intensity and the experimental error at the momentum transfer s_j, respectively. Representative models were generated by averaging of multiple SA runs using DAMAVER [63, 64].

Spectroscopic methods

All experiments were performed at 25 °C in 50 mM sodium phosphate and 100 mM NaCl, pH 7.5. Protein concentration measurements were performed with a JASCO V550 spectrophotometer. The concentrations of RCM-T1 and RCM-α-lactalbumin were determined by using an absorption coefficent of $\varepsilon_{278} = 21060$ M^{-1} cm^{-1} and $\varepsilon_{278} = 28340$ M^{-1} cm^{-1}. The molar absorption coefficents (ε_{280}) of tSlyD (5960 M^{-1} cm^{-1}), tSlyDΔIF (4470 M^{-1} cm^{-1}), tIF (1490 M^{-1} cm^{-1}), and the Tat-signalpeptide (1-27) (6990 M^{-1} cm^{-1}) were calculated as described [65]. GdmCl induced unfolding transitions monitored by fluorescence were recorded with a JASCO FP6500 fluorescence spectrometer. A fluorescence spectrum was recorded for each data point from 300 to 400 nm after excitation at 276 nm. Experimental data were analyzed at a single wavelength according to a two-state model by non-linear least-squares fit with proportional weighting to obtain the Gibbs free energy of denaturation ΔG_D as a function of the GdmCl concentration using GraFit 5 (Erithacus Software, Staines, UK).

To determine the dissociation constants of tSlyD (+ variants) with substrates, Trp fluorescence increase upon binding was measured at 335 nm (for the Tat-signalpeptide) and at 343 nm (for

RCM-α-lactalbumin) after excition at 295 nm. The tryptophan fluorescence intensity was corrected for buffer fluorescence and dilution. The resulting binding isotherms were analyzed as described [66].

Folding experiments

RCM-T1 was unfolded by incubating the protein in 0.1 M Tris-HCl (pH 8.0) at 15 °C for at least 1 h. Refolding at 15 °C was initiated by a 40-fold dilution of the unfolded protein to final conditions of 2.0 M NaCl and the desired concentrations of the prolyl isomerase and RCM-T1 in the same buffer. The folding reaction was followed by the increase in protein fluorescence at 320 nm after excitation at 268 nm. At 2.0 M NaCl, slow folding of RCM-T1 was a monoexponential process, and its rate constant was determined using Grafit 5.

NMR

All NMR spectra were aquired with a Bruker Avance 600 and 800 MHz spectrometer in 50 mM sodium phosphate buffer, 100 mM NaCl, pH 7.5, containing 10 % ^2H_2O at 25 °C. For backbone assignment trHNCA, trHNCACB, and trHN(CO)CACB were measured with a 2.2 mM $^{15}N/^{13}C$ labelled sample. Spectra were processed using NMRpipe [67] and analyzed with NMRView [68]. The H/D exchange reaction was started by buffer exchange of ^{15}N-labelled tSlyD in ^2H_2O buffer, containing 50 mM sodium phosphate, 100 mM NaCl at pD 7.1 (pH meter reading). A series of 110 2D $^1H/^{15}N$-HSQC spectra was recorded during an exchange time of 73 hours. Protection factors (P) were derived from $P = k_{int}/k_{ex}$ (assuming EX2 regime), where k_{int} is the intrinsic exchange rate constant from peptide models [69] and k_{ex} is the observed rate constant obtained by fitting a single exponential function to the volume decay of amide cross peaks. The Gibbs free energy of complete unfolding for tSlyD of 35.1 ± 3.0 kJ/mol (see Table 1) converts into a protection factor of $3 \cdot 10^6$ by

$\Delta G_U = -RT \ln(1/P)$ which is in good aggreement with the most protected amide protons observed.

The tSlyD titration experiment was carried out by successive addition of aliquots of a stock solution RCM-α-lactalbumin in the same buffer as the protein. Complex formation was monitored by recording a 2D $^1H/^{15}N$-HSQC spectrum after each titration step. The mean weighted difference in the chemical shifts for most of the residues were calculated according to following equation: $\Delta\delta_{MW}(^1H,^{15}N) = [(\Delta\delta(^1H)^2 + 1/25 \cdot \Delta\delta(^{15}N)^2)/2]^{0.5}$ [70].

Protein Data Bank accession code

The data are deposited in the RCSB Protein Data Bank and are available under accession codes 3CGM and 3CGN.

Acknowledgement

We thank Paul Rösch for NMR spectrometer time at 800 MHz, Kristian Schweimer for measurement of the assignment spectra, Christian Lange for help with ITC measurements, Caroline Haupt for chaperone assay and Efstratios Mylonas for assistance with SAXS measurements. This research was supported by grants from the Deutsche Forschungsgemeinschaft (Ba 1821/3-1,2 and GRK 1026 "Conformational transitions in macromolecular interactions") and the excellence initiative of the state Sachsen-Anhalt.

Table 1. Data collection and refinement statistics of tSlyD structures

Data collection	tSlyD (crystal form A)	tSlyD (crystal form B)
Wavelength (Å)	$\lambda = 0.91840$	$\lambda = 0.91840$
Space group	P6(3)22	F222
a (Å)	121.780	81.18
b (Å)	121.780	85.09
c (Å)	76.260	92.22
Max. resolution (Å)	2.41	2.70
R_{merge}[a]	9.3 (58.2)[b]	5.4 (52.6)
% complete	96.7 (91.8)	90.6 (78.2)
No. of reflections	55063	21311
No. of unique reflections	13468	4147
$<I/\sigma(I)>$	8.75 (2.46)	16.83 (2.80)
Refinement		
Resolution range (Å)	28.27–2.41 (2.47-2.41)	20.00–2.70 (2.70-2.87)
Completeness (working + test) (%)	100.0 (100.0)	90.6 (80.2)
No. of reflections ($F>0$)	10983 (969)	4141 (676)
Wilson B (Å^2)	75.3	79.5
R_{cryst}[c] (%)	18.70 (25.70)	20.90 (29.10)
R_{free}[d] (%)	22.57 (29.20)	25.70 (35.55)
No. of non-hydrogen atoms		
Protein	1253	1163
Water	98	21
SCN	18	-
Glycerol	12	-
Ni	1	-
SO_4^{2-}	-	5
R.m.s.d. from ideality		
Bond lengths (Å)	0.015	0.007
Bond angles (°)	1.62	1.30
Dihedral angles (°)	25.5	25.3
Improper angles (°)	0.97	0.95
Average B-factor (Å^2)		
Protein atoms	41.37	85.19
Main chain	40.41	84.54
Water	45.70	60.67
SCN, Glycerol	62.07	
Ni^{2+}	35.55	
SO_4^{2-}	-	54.93

[a] $R_{merge} = \sum |I - <I>| / \sum <I>$.

[b] Values in parentheses correspond to the highest resolution shell (2.47–2.41 Å).

[c] $R_{cryst} = \sum ||Fobs| - |Fcalc|| / \sum |Fobs|$.

[d] R_{free} is calculated as R_{cryst} for a test set comprising 7.5% reflections not used in the refinement.

Table 2. Stabilities and catalytic efficiencies of tSlyD and variants

Stability data were derived at 15 °C[f] or 25 °C[a-c], pH 7.5 from GdmCl[a-c] or urea[f] induced unfolding transition. Transition midpoints ($[D_{½}]$), cooperativity values (m) and the Gibbs free energies of denaturation at 0 M denaturant ($\Delta G_D(H_2O)$) are given. Activity data result from catalysis of refolding of RCM-T1 at 15 °C in 0.1 M Tris-HCl pH 8.0, 2.0 M NaCl. The data for Thermus BP12 refer to the variant with the C22A mutation in the FKBP12 domain; n.d. not determined. [g] taken from [38].

construct	$[D_{½}]$ (M)	m (kJ mol^{-1} M^{-1})	$\Delta G_D(H_2O)$ (kJ/mol)	k_{cat}/K_M (M^{-1} s^{-1})
tSlyD[a]	3.1	11.4 ± 0.5	35.1 ± 3.0	0.29 · 10^6
tSlyDΔIF[b]	3.7	8.7 ± 0.5	32.0 ± 1.2	0.25 · 10^4
tIF[c]	1.0	4.3 ± 0.3	4.3 ± 0.3	no acitivity
FKBP12[d]	n.d.	n.d.	n.d.	1.46 · 10^4
FKBP12 C22A[e]	n.d.	n.d.	n.d.	1.50 · 10^4 [g]
Thermus BP12[f]	2.2	5.4 ± 0.3	11.8 ± 0.5	2.71 · 10^6

Table 3. Binding properties of tSlyD and variants to various substrates

Binding data were derived at 25 °C, pH 7.5. Binding affinities and stoichiometries (given as ratio (prolyl isomerase/substrate) derived from analyzing the binding isotherms assuming the same binding affinity at multiple binding sites. [a]Stoichiometry set constant during fitting procedure.

construct	method	substrate	K_D (µM)	stoichiometry
tSlyD[a]	fluorescence	RCM-α-lactalbumin	0.51 ± 0.08	2
tSlyD	NMR	RCM-α-lactalbumin	n.d.	1.9 ± 0.2
tSlyD	fluorescence	Tat-signalpeptide	$51 \pm 21 \times 10^{-3}$	1.05 ± 0.03
tSlyD	ITC	Tat-signalpeptide	$113 \pm 20 \times 10^{-3}$	0.99 ± 0.02
tSlyD	NMR	Tat-signalpeptide	n.d.	1.0 ± 0.05
tSlyDΔIF	fluorescence	RCM-α-lactalbumin	no binding	-
tSlyDΔIF	fluorescence	Tat-signalpeptide	no binding	-
tIF[a]	fluorescence	RCM-α-lactalbumin	1.1 ± 0.1	2

References

1. Schindler, T., Herrler, M., Marahiel, M.A., and Schmid, F.X. (1995). Extremely rapid folding in the absence of intermediates: the cold-shock protein from Bacillus subtilis. Nat. Struct. Biol. *2*, 663-673.
2. Huang, G.S., and Oas, T.G. (1995). Submillisecond folding of monomeric lambda repressor. Proc Natl Acad Sci U S A *92*, 6878-6882.
3. Eaton, W.A., Munoz, V., Hagen, S.J., Jas, G.S., Lapidus, L.J., Henry, E.R., and Hofrichter, J. (2000). Fast kinetics and mechanisms in protein folding. Annu Rev Biophys Biomol Struct *29*, 327-359.
4. Ferguson, N., and Fersht, A.R. (2003). Early events in protein folding. Curr Opin Struct Biol *13*, 75-81.
5. Brandts, J.F., Halvorson, H.R., and Brennan, M. (1975). Consideration of the possibility that the slow step in protein denaturation reactions is due to cis-trans isomerism of proline residues. Biochemistry *14*, 4953-4963.
6. Schmid, F.X. (1982). Proline isomerization in unfolded ribonuclease A. European Journal of Biochemistry *128*, 77-80.
7. Bardwell, J.C.A. (1994). Building bridges: Disulphide bond formation in the cell. Molecular Microbiology *14*, 199-205.
8. Kadokura, H., Tian, H., Zander, T., Bardwell, J.C., and Beckwith, J. (2004). Snapshots of DsbA in action: detection of proteins in the process of oxidative folding. Science *303*, 534-537.
9. Dobson, C.M. (2003). Protein folding and misfolding. Nature *426*, 884-890.
10. Fuchs, A., Seiderer, C., and Seckler, R. (1991). Invitro Folding Pathway of Phage-P22 Tailspike Protein. Biochemistry *30*, 6598-6604.
11. Dobson, C.M. (2003). Protein folding and misfolding. Nature *426*, 884-890.
12. Dobson, C.M. (2004). Principles of protein folding, misfolding and aggregation. Semin Cell Dev Biol *15*, 3-16.
13. Collet, J.F., Riemer, J., Bader, M.W., and Bardwell, J.C. (2002). Reconstitution of a disulfide isomerization system. J Biol Chem *277*, 26886-26892.
14. Fischer, G., Wittmann-Liebold, B., Lang, K., Kiefhaber, T., and Schmid, F.X. (1989). Cyclophilin and peptidyl-prolyl cis-trans isomerase are probably identical proteins. Nature *337*, 476-478.
15. Bukau, B., Deuerling, E., Pfund, C., and Craig, E.A. (2000). Getting newly synthesized proteins into shape. Cell *101*, 119-122.
16. Young, J.C., Agashe, V.R., Siegers, K., and Hartl, F.U. (2004). Pathways of chaperone-mediated protein folding in the cytosol. Nat Rev Mol Cell Biol *5*, 781-791.
17. Walter, S., and Buchner, J. (2002). Molecular chaperones--cellular machines for protein folding. Angew Chem Int Ed Engl *41*, 1098-1113.
18. Beissinger, M., and Buchner, J. (1998). How chaperones fold proteins. Biol Chem *379*, 245-259.
19. Ramm, K., and Plückthun, A. (2000). The periplasmic Escherichia coli peptidylprolyl cis,trans-isomerase FkpA. II. Isomerase-independent chaperone activity in vitro. J Biol Chem *275*, 17106-17113.
20. Ramm, K., and Plückthun, A. (2001). High enzymatic activity and chaperone function are mechanistically related features of the dimeric E. coli peptidyl-prolyl-isomerase FkpA. J. Mol. Biol. *310*, 485-498.
21. Saul, F.A., Arie, J.P., Vulliez-le Normand, B., Kahn, R., Betton, J.M., and Bentley, G.A. (2004). Structural and functional studies of FkpA from Escherichia coli, a cis/trans peptidyl-prolyl isomerase with chaperone activity. J. Mol. Biol. *335*, 595-608.
22. Hesterkamp, T., and Bukau, B. (1996). Identification of the prolyl isomerase domain of Escherichia coli trigger factor. FEBS Lett. *385*, 67-71.
23. Stoller, G., Rücknagel, K.P., Nierhaus, K., Schmid, F.X., Fischer, G., and Rahfeld, J.-U. (1995). Identification of the peptidyl-prolyl cis/trans isomerase bound to the Escherichia coli ribosome as the trigger factor. EMBO Journal *14*, 4939-4948.
24. Scholz, C., Stoller, G., Zarnt, T., Fischer, G., and Schmid, F.X. (1997). Cooperation of enzymatic and chaperone functions of trigger factor in the catalysis of protein folding. EMBO Journal *16*, 54-58.
25. Rouviere, P.E., and Gross, C.A. (1996). SurA, a periplasmic protein with peptidyl-prolyl isomerase activity, participates in the assembly of outer membrane porins. Gene Develop. *10*, 3170-3182.
26. Scholz, C., Schaarschmidt, P., Engel, A.M., Andres, H., Schmitt, U., Faatz, E., Balbach, J., and Schmid, F.X. (2004). Functional solubilization of aggregation-prone HIV envelope proteins by covalent fusion with chaperone modules. J. Mol. Biol., 1229-1241.
27. Scholz, C., Eckert, B., Hagn, F., Schaarschmidt, P., Balbach, J., and Schmid, F.X. (2006). SlyD proteins from different species exhibit high prolyl isomerase and chaperone activities. Biochemistry *45*, 20-33.
28. Heras, B., Edeling, M.A., Schirra, H.J., Raina, S., and Martin, J.L. (2004). Crystal structures of the DsbG disulfide isomerase reveal an unstable disulfide. Proc Natl Acad Sci U S A *101*, 8876-8881.

29. McCarthy, A.A., Haebel, P.W., Torronen, A., Rybin, V., Baker, E.N., and Metcalf, P. (2000). Crystal structure of the protein disulfide bond isomerase, DsbC, from Escherichia coli. Nat Struct Biol *7*, 196-199.
30. Hottenrott, S., Schumann, T., Plückthun, A., Fischer, G., and Rahfeld, J.U. (1997). The Escherichia coli SlyD is a metal ion-regulated peptidyl-prolyl cis/trans-isomerase. J. Biol. Chem. *272*, 15697-15701.
31. Fischer, G. (1994). Peptidyl-prolyl cis/trans isomerases and their effectors. Angew.Chem.Int.Ed. *33*, 1415-1436.
32. Gothel, S.F., and Marahiel, M.A. (1999). Peptidyl-prolyl cis-trans isomerases, a superfamily of ubiquitous folding catalysts. Cell Mol Life Sci *55*, 423-436.
33. Weiwad, M., Werner, A., Rucknagel, P., Schierhorn, A., Kullertz, G., and Fischer, G. (2004). Catalysis of proline-directed protein phosphorylation by peptidyl-prolyl cis/trans isomerases. J Mol Biol *339*, 635-646.
34. Fischer, G., and Aumuller, T. (2003). Regulation of peptide bond cis/trans isomerization by enzyme catalysis and its implication in physiological processes. Rev Physiol Biochem Pharmacol *148*, 105-150.
35. Schreiber, S.L., and Crabtree, G.R. (1992). The Mechanism of Action of Cyclosporin-A and FK506. Immunol.Today *13*, 136-142.
36. Galat, A., and Metcalfe, S.M. (1995). Peptidylproline cis/trans isomerases. Prog Biophys Mol Biol *63*, 67-118.
37. Ikura, T., and Ito, N. (2007). Requirements for peptidyl-prolyl isomerization activity: A comprehensive mutational analysis of the substrate-binding cavity of FK506-binding protein 12. Protein Sci *16*, 2618-2625.
38. Knappe, T.A., Eckert, B., Schaarschmidt, P., Scholz, C., and Schmid, F.X. (2007). Insertion of a chaperone domain converts FKBP12 into a powerful catalyst of protein folding. J Mol Biol *368*, 1458-1468.
39. Bernhardt, T.G., Roof, W.D., and Young, R. (2002). The Escherichia coli FKBP-type PPIase SlyD is required for the stabilization of the E lysis protein of bacteriophage phi X174. Mol. Microbiol. *45*, 99-108.
40. Roof, W.D., and Young, R. (1995). Phi X174 lysis requires slyD, a host gene which is related to the FKBP family of peptidyl-prolyl cis-trans isomerases. FEMS Microbiol Rev *17*, 213-218.
41. Roof, W.D., Fang, H.Q., Young, K.D., Sun, J., and Young, R. (1997). Mutational analysis of slyD, an Escherichia coli gene encoding a protein of the FKBP immunophilin family. Mol Microbiol *25*, 1031-1046.
42. Roof, W.D., Horne, S.M., Young, K.D., and Young, R. (1994). slyD, a host gene required for phi X174 lysis, is related to the FK506-binding protein family of peptidyl-prolyl cis-trans-isomerases. J Biol Chem *269*, 2902-2910.
43. Mendel, S., Holbourn, J.M., Schouten, J.A., and Bugg, T.D. (2006). Interaction of the transmembrane domain of lysis protein E from bacteriophage phiX174 with bacterial translocase MraY and peptidyl-prolyl isomerase SlyD. Microbiology *152*, 2959-2967.
44. Butland, G., Zhang, J.W., Yang, W., Sheung, A., Wong, P., Greenblatt, J.F., Emili, A., and Zamble, D.B. (2006). Interactions of the Escherichia coli hydrogenase biosynthetic proteins: HybG complex formation. FEBS Lett *580*, 677-681.
45. Zhang, J.W., Butland, G., Greenblatt, J.F., Emili, A., and Zamble, D.B. (2005). A role for SlyD in the Escherichia coli hydrogenase biosynthetic pathway. J Biol Chem *280*, 4360-4366.
46. Graubner, W., Schierhorn, A., and Bruser, T. (2007). DnaK plays a pivotal role in Tat targeting of CueO and functions beside SlyD as a general Tat signal binding chaperone. J Biol Chem *282*, 7116-7124.
47. Han, K.Y., Song, J.A., Ahn, K.Y., Park, J.S., Seo, H.S., and Lee, J. (2007). Solubilization of aggregation-prone heterologous proteins by covalent fusion of stress-responsive Escherichia coli protein, SlyD. Protein Eng Des Sel *20*, 543-549.
48. Wagner, S., Baars, L., Ytterberg, A.J., Klussmeier, A., Wagner, C.S., Nord, O., Nygren, P.A., van Wijk, K.J., and de Gier, J.W. (2007). Consequences of membrane protein overexpression in Escherichia coli. Mol Cell Proteomics *6*, 1527-1550.
49. Scholz, C., Thirault, L., Schaarschmidt, P., Zarnt, T., Faatz, E., Engel, A.M., Upmeier, B., Bollhagen, R., Eckert, B., and Schmid, F.X. (2008). Chaperone-Aided in Vitro Renaturation of an Engineered E1 Envelope Protein for Detection of Anti-Rubella Virus IgG Antibodies. Biochemistry.
50. Suzuki, R., Nagata, K., Yumoto, F., Kawakami, M., Nemoto, N., Furutani, M., Adachi, K., Maruyama, T., and Tanokura, M. (2003). Three-dimensional solution structure of an archaeal FKBP with a dual function of peptidyl prolyl cis-trans isomerase and chaperone-like activities. J Mol Biol *328*, 1149-1160.
51. Koradi, R., Billeter, M., and Wüthrich, K. (1996). MOLMOL: a program for display and analysis of macromolecular structures. J Mol Graph *14*, 51-55.

52. Svergun, D.I. (1999). Restoring low resolution structure of biological macromolecules from solution scattering using simulated annealing. Biophys J *76*, 2879-2886.
53. Scheibel, T., Weikl, T., and Buchner, J. (1998). Two chaperone sites in Hsp90 differing in substrate specificity and ATP dependence. Proc Natl Acad Sci U S A *95*, 1495-1499.
54. Hu, K., Galius, V., and Pervushin, K. (2006). Structural plasticity of peptidyl-prolyl isomerase sFkpA is a key to its chaperone function as revealed by solution NMR. Biochemistry *45*, 11983-11991.
55. Mücke, M., and Schmid, F.X. (1994). Folding mechanism of ribonuclease T1 in the absence of the disulfide bonds. Biochemistry *33*, 14608-14619.
56. Emsley, P., and Cowtan, K. (2004). Coot: model-building tools for molecular graphics. Acta Crystallogr D Biol Crystallogr *60*, 2126-2132.
57. Murshudov, G.N., Vagin, A.A., and Dodson, E.J. (1997). Refinement of macromolecular structures by the maximum-likelihood method. Acta Crystallogr D Biol Crystallogr *53*, 240-255.
58. (1994). The CCP4 suite: programs for protein crystallography. Acta Crystallogr D Biol Crystallogr *50*, 760-763.
59. Roessle, M.W., R. Klaering, et al. (2007). Upgrade of the Small Angle X-ray scattering Beamline X33 at the EMBL Hamburg. J. Appl. Crystallogr, in press.
60. Volkov, V.V., Konarev, P.V., Sokolova, A.V., Koch, M.H.J., and Svergun, D.I. (2003). Primus: A Windows PC-based system for small-angle x-ray scattering data analysis. J Appl Crystallogr *36*, 1277-1282.
61. Guinier, A. (1939). La diffraction des rayons X aux tres petits angles: application a l'etude de phenomenes ultramicroscopiques. Ann. Phys. *12*, 161-237.
62. Svergun, D.I. (1992). Determination of the regularization parameter in indirect-transform methods using perceptual criteria. J. Appl. Crystallogr. *25*, 495-503.
63. Kozin, M.B., and Svergun, D.I. (2001). Automated matching of high- and low-resolution structural models. J. Appl. Crystallogr. *34*, 33-41.
64. Volkov, V.V., and Svergun, D.I. (2003). Uniqueness of ab initio shape determination in small-angle scatering. J. Appl. Crystallogr. *36*, 860-864.
65. Gill, S.C., and von Hippel, P.H. (1989). Calculation of protein extinction coefficients from amino acid sequence data. Anal. Biochem. *182*, 319-326.
66. Zeeb, M., and Balbach, J. (2003). Single-stranded DNA binding of the cold shock protein CspB from Bacillus subtilis: NMR mapping and mutational characterisation. Protein Sci. *12*, 112-123.
67. Delaglio, F., Grzesiek, S., Vuister, G.W., Zhu, G., Pfeifer, J., and Bax, A. (1995). NMRPipe: a multidimensional spectral processing system based on UNIX pipes. J. Biomol. NMR *6*, 277-293.
68. Johnson, B.A., and Blevins, R.A. (1994). NMRView: A computer program for visualization and analysis of NMR data. Journal of Biomolecular NMR *4*, 603-614.
69. Bai, Y.W., Milne, J.S., Mayne, L., and Englander, S.W. (1993). Primary Structure Effects on Peptide Group Hydrogen Exchange. Proteins: Structure Function and Genetics *17*, 75-86.
70. Grzesiek, S., Stahl, S.J., Wingfield, P.T., and Bax, A. (1996). The CD4 determinant for downregulation by HIV-1 Nef directly binds to Nef. Mapping of the Nef binding surface by NMR. Biochemistry *35*, 10256-10261.
71. DeLano, W. (2003). The PyMOL Molecular Graphics System. DeLano Scientific LLC, San Carlos, CA, USA. http://www.pymol.org.

Subproject D

SI Information

SI Figure 8. Densitiy map of 6-fold coordinated Ni^{2+} ion found in tSlyD structure derived from crystal form A. A C-terminal histidine motif of tSlyD (histidine 145, 147 and 149), which is higly conserved among the SlyD family, provides three coordination sites, while two histidine residues of the own His-tag (Histidine 155 and 157) and histidine 153 of a neighbouring molecule complement the 6-fold coordination. The figure was created using PyMOL [71].

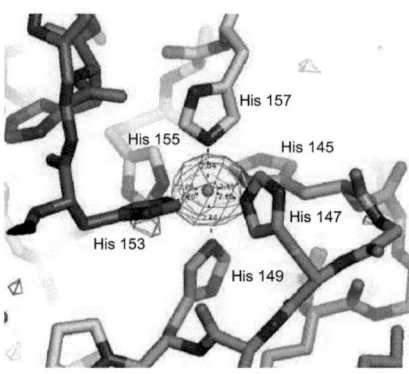

SI Figure 9. 2D ^{1}H-^{15}N-TROSY-HSQC spectrum of 2.0 mM ^{15}N-tSlyD at 25 °C in 90%/ 10% ^{1}H$_2$O/^{2}H$_2$O, pH 7.5. The assigned cross-peaks of the amide backbone are labeled using the one-letter amino acid code and the sequence position. Boxes indicate resonance signals, which show cross-peak intensities below the plotted contour level.

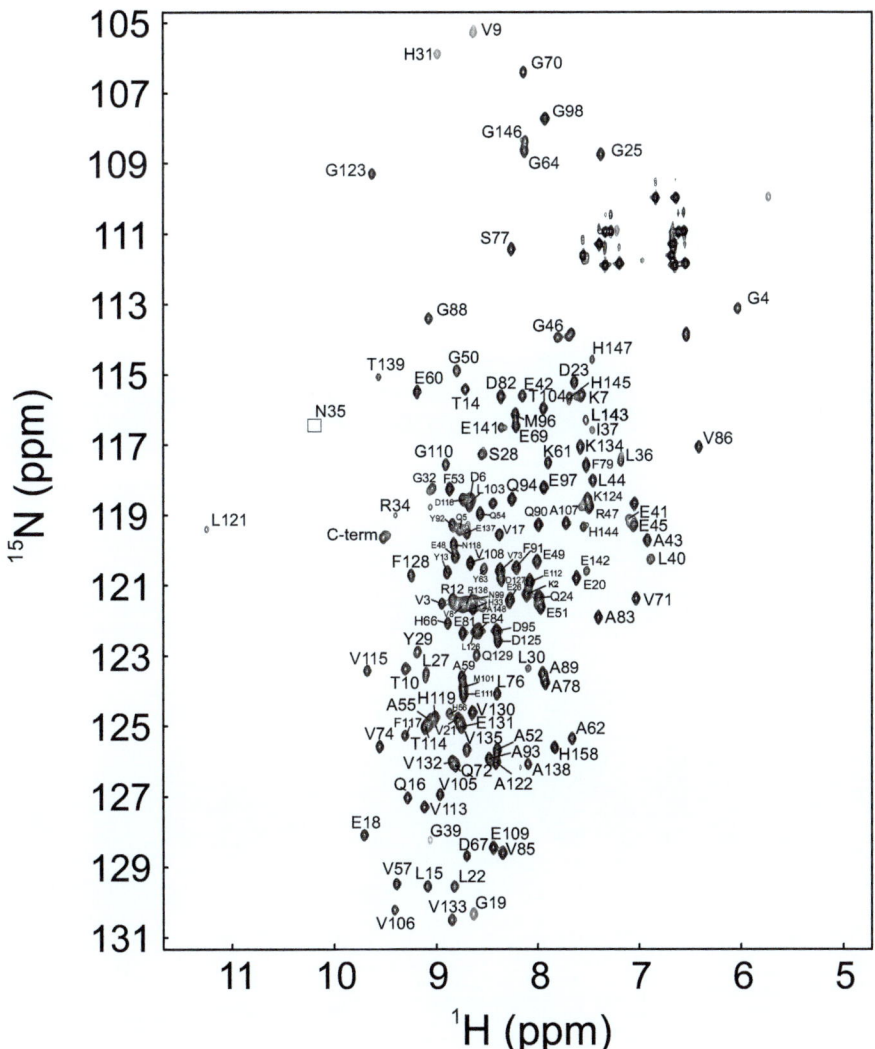

SI Figure 10. (a) Linearity of the Guinier plots of SAXS data for free tSlyD and a tSlyD-Tat-signalpeptide complex together with the agreement of their molecular masses (determined from the extrapolated I(0) values) suggest that the solutions are homogeneous. Colour code as in Fig. 3. (b) Kratky plots of SAXS data for free tSlyD and a tSlyD/Tat-signalpeptide complex normalized to the maximum value of unity. Colour code as in Fig. 3. The maximum at small angles indicates the presence of folded domains. *Ab initio* low resolution structure models of tSlyD (c) and the tSlyD-Tat-signalpeptide complex (d) calculated from the SAXS pattern. The balls represent the dummy atoms in the simulated annealing procedure to restore the models.

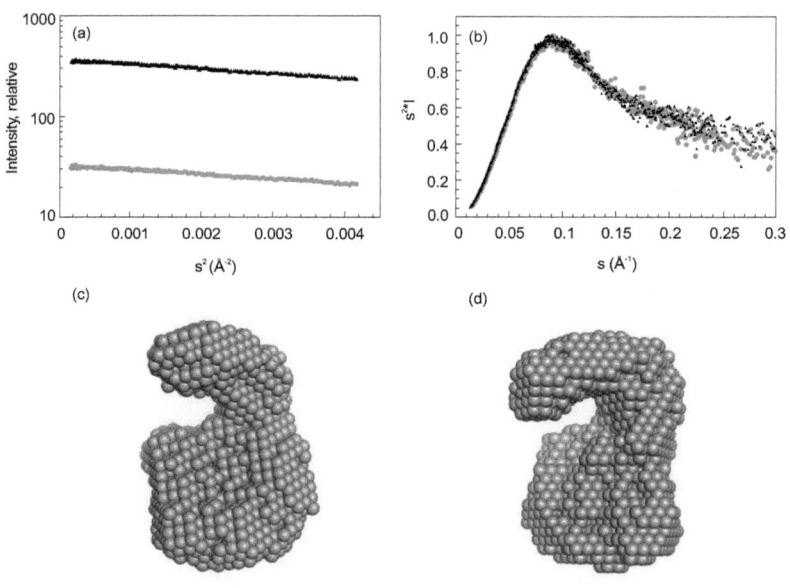

SI Figure 11. Chaperone properties of tSlyD and the isolated domains. Aggregation of Insulin after reduction of the disulfide bridges with DTT at 25 °C (○) in the absence and (□) in the presence of 4.5 µM tSlyD, (●) 10 µM tSlyDΔIF, and (▲) 20 µM tIF domain. 45 µM insulin was reduced with 10 mM DTT. The increase of light scattering upon aggregation was monitored at 450 nm as a function of time.

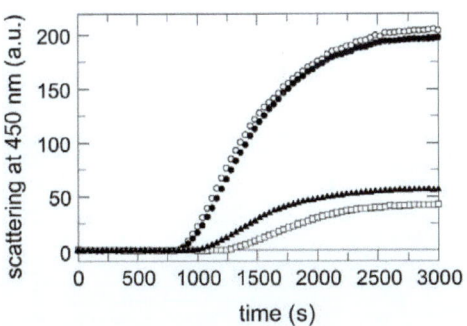

SI Figure 12. Catalytic efficiencies of refolding of 0.1 µM RCM-T1 in the absence and presence of prolyl isomerases. The measured rate constants k_{app} of RCM-T1 folding are shown as function of Thermus BP12 (●), tSlyD (△), FKBP12 (□), tSlyDΔIF (■) and the tIF domain (▲). The k_{cat}/K_M values derived from the slopes are given in Table 2. Refolding kinetics were measured at 15 °C in 0.1 M Tris-HCl (pH 8.0), 2.0 M NaCl, by the change in fluorescence at 320 nm after excitation at 268 nm.

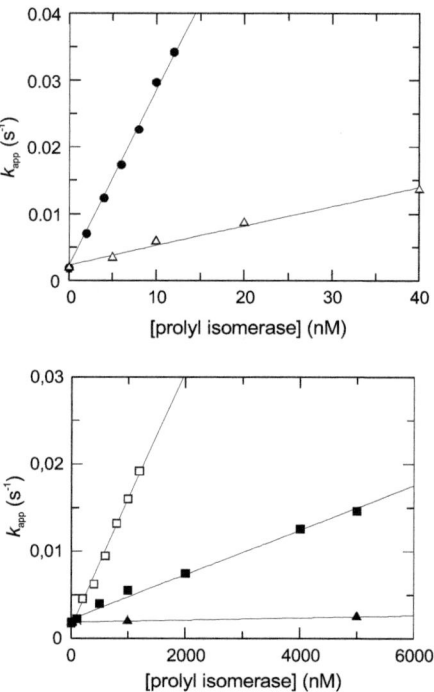

7.5 Subproject E

Subproject E

Subproject E

Structure and dynamics of Helix-0 of the N-BAR domain in Lipid Micelles and Bilayers

Christian Löw[1], Ulrich Weininger[1], Hwankyu Lee[2], Kristian Schweimer[3], Ines Neundorf[4], Annette G. Beck-Sickinger[4], Richard W. Pastor[2], Jochen Balbach[*,1,5]

[1] Institut für Physik, Biophysik, Martin-Luther-Universität Halle-Wittenberg, D-06120 Halle (Saale), Germany

[2] Laboratory of Computational Biology, National Heart, Lung, and Blood Institute, National Institutes of Health, Bethesda, MD 20892, USA

[3] Lehrstuhl Biopolymere, Universität Bayreuth, Universitätsstrasse 30, 95447 Bayreuth, Germany

[4] Institut für Biochemie, Fakultät für Biowissenschaften, Pharmazie und Psychologie, Universität Leipzig, Brüderstr. 34, D-04103 Leipzig, Germany

[5] Mitteldeutsches Zentrum für Struktur und Dynamik der Proteine (MZP), Martin-Luther-Universität Halle-Wittenberg, Germany

Runnig title: Structure of Helix-0 of the N-BAR domain

*Correspondence should be addressed to
Jochen Balbach Tel.: ++49 345 55 25353
Institut für Physik, Fachgruppe Biophysik Fax: ++49 345 55 27383
Martin-Luther-Universität Halle-Wittenberg E-Mail: jochen.balbach@physik.uni-halle.de
D-06120 Halle(Saale), Germany

Abstract

BAR domains (Bin/Amphiphysin/Rvs-homology) generate and sense membrane curvature by binding the negatively charged membrane to their positively charged concave surfaces. N-BAR domains contain an N-terminal extension (helix-0) predicted to form an amphipathic helix upon membrane binding. We determined the NMR structure and nano-to-picosecond dynamics of helix-0 of the human Bin1/Amphiphysin II BAR domain in SDS and DPC micelles. Molecular dynamic simulations of this 34 amino acid peptide revealed electrostatic and hydrophobic interactions with the detergent molecules, which induce helical structure formation from residues 8-10 towards the C-terminus. The orientation in the micelles was experimentally confirmed by backbone amide proton exchange. Both simulation and experiment indicate that the N-terminal region is disordered, and the peptide curves to adopt the micelle shape. Deletion of helix-0 reduces tubulation of liposomes by the BAR domain, whereas the helix-0 peptide itself was fusogenic. These findings support models for membrane curving by BAR domains, where helix-0 increases the binding affinity to the membrane and enhances curvature generation.

Abbreviations: DPC, dodecyl phospho choline; FRET, fluorescence resonance energy transfer; MEXICO, measurements of exchange rates in isotopically labeled compounds; OG, n-Octyl-β-D-glucopyranoside; SDS, sodium dodecyl sulphate

Keywords: membrane curvature, NMR structure, MD simulations, liposomes

Introduction

Proteins play fundamental roles in modulating the structure of lipid bilayers. Processes such as membrane fusion, budding or tubulation are associated with changes in membrane curvature. The banana shaped BAR (Bin/Amphiphysin/Rvs-homology) domains have been identified throughout eukarya as regulators of membrane remodeling processes. They sense and curve membranes, and participate in numerous cytoskeletal and nuclear processes, such as clathrin–mediated endocytosis or organization of the T-tuble network in the muscle (1-10). Point mutations found in centronuclear myopathy patients in the BAR domain of Bin1 causes a dysfunction of the latter process (11).

The crystal structures of the human and drosophila amphiphysin BAR domain (12,13) reveal a crescent shaped homodimer with a positively charged concave surface. This suggests that driving and/or sensing curvatures of membranes by BAR domains occurs by binding of negatively charged membranes to this positively charged surface (13). Some BAR domains (denoted N-BAR) contain an N-terminal extension with amphipathic character which is predicted to undergo a random coil to helix transition by binding to the membrane (13). This extension, termed helix-0, shows no electron density in the crystal structure (14). *In vitro*, BAR domains can induce curvature in liposomes, resulting in narrow tubes (`tubulation`) (15). Recent experimental and theoretical studies (16) with N-BAR domains indicate that helix-0 embeds in the lipid bilayer and strongly increases the ability to tubulate liposomes. The insertion of amphipathic helices into hydrophobic phases of the bilayer has been proposed to be a general mechanism for curvature generation during vesicle budding as shown in amphiphysin (13) and further examples (17-19). Experimental evidence for structure induction and insertion of the amphipathic helix has been derived from circular dichroism (CD) and electron paramagnetic resonance spectroscopy (EPR) (13,14,20,21). There are currently, three candidate curvature-generating mechanisms: the local spontaneous curvature, the bilayer-coupling, and the scaffolding (22). The scaffold mechanism assumes that the intrinsic curvature of the BAR domain forces the membrane shape, as opposed to a deformation of the lipid bilayer by a shallow (spontaneous curving) or deep (bilayer-coupling) insertion of an amphipathic helix.

To obtain further insights into the predicted N-terminal amphipathic helix of N-BAR domains, we studied helix-0 of the human Bin1/Amphiphysin II BAR domain (N-BAR) in detergent and lipid environments by high resolution NMR spectroscopy and molecular dynamics (MD) simulations. Structure calculation, dynamic measurements, and a fast amide proton exchange confirmed the earlier proposed amphipathic character of the induced helix but also revealed a disordered N-terminal part of the amphipathic helix, which is highly flexible and exposed to the solvent. The balance of electrostatic and hydrophobic interactions is considered. Lastly, a tubulation assay of liposomes analyzed by electron microscopy or fluorescence resonance energy transfer shows that the isolated N-BAR peptide is fusogenic.

NMR spectroscopy
NMR-spectra were acquired with a Bruker Avance 800, Bruker Avance 700 equipped with a cryoprobe, and a Bruker Avance II 600 spectrometer in 20 mM sodium phosphate buffer, pH 7.4, containing 10 % ^2H_2O at 25 °C except for the free N-BAR peptide where experiments were carried out at 15 °C and the extended N-BAR peptide (1-44 residues) at pH 6.0 and 15 °C. The N-BAR peptide in SDS and DPC micelles as well as the unbound form were assigned by 3D-^{15}N-TOCSY-HSQC and 3D-^{15}N-NOESY-HSQC (see SI Table 3). For structure calculation of SDS- and DPC-bound N-BAR peptide, an additional 2D-NOESY spectrum and for the SDS bound form a 3D-HNHA spectrum were recorded. For further investigations a ^{15}N heteronuclear NOE and a MEXICO proton-exchange experiment (24) were performed. The ordinate in Fig. 5 corresponds to the NMR cross peak intensity at the respective exchange time divided by intensity in a reference experiment. Spectra were processed with NMRpipe (25) and analyzed with NMRView (26).

Structure calculation
Distance restraints were obtained from 3D-^{15}N-NOESY-HSQC and 2D-NOESY and used ambiguously for structure calculation with ARIA (27). Backbone dihedral restraints were calculated from chemical shifts using TALOS. It should be noted that the random coil values of TALOS are not optimized for the micellar environment. ARIA runs with and without TALOS restrains however gave the same overall topology and curvature but reduced r.m.s.d. values and therefore TALOS was included in the final run. Structure geometry was analyzed with PROCHECK (28).

Computational methods
All simulations and analyses were performed using the GROMACS 3.3.1 simulation package (29,30). Coordinates of the SDS molecule were generated using the PRODRG2 server (31). The united-parameter set for lipids was downloaded from http://moose.bio.ucalgary.ca, and charges of SDS head group were set equal to those in the CHARMM force field (32); OPLS parameters (33) were used for the peptide. Five initial conditions were constructed (Table 1 and Figure 6). "SDS1", "SDS2", and "SDS3" contained 75 lipids (the experimentally established aggregation number for SDS micelles (34). "SDS4" contained 40 lipids to investigate effects of micelle size, and "DPC" contained 65 monomers of DPC. Initial coordinates for the N-BAR peptide were those of the lowest energy structure from the NMR ensemble, which is 76 % α-helical. The peptides were placed in different orientations and positions with respect to the micelle as specified in Table 1. Approximately 16,000 TIP4P water molecules were placed around a mixture of the peptide and micelle to a thickness of 1 nm, forming a periodic box of size $8 \times 8 \times 8$ nm^3. Na$^+$ ions were added to neutralize charges from the SDS molecules. For DPC, the equilibrated 65-surfactant DPC micelle was downloaded from http://moose.bio.ucalgary.ca, and the same procedures were performed with 5 Cl$^-$ ions added to neutralize the peptide. A pressure of 1 bar and a temperature of 298 K were maintained in an NPT ensemble with the weak coupling method (35). After energy minimization, equilibration runs were performed for 60 ns without any experimental distance restrain, and the coordinates were saved every ps for analysis. Secondary structure of the peptide was calculated using the DSSP program (36).

Protein structure accession number
The coordinates of the structure of the N-BAR peptide in DPC and SDS micelles have been deposited in the Protein Databank under accession No. 2RND and 2RMY.

Materials and Methods

Expression and purification of the N-BAR and BAR Domain and the N-BAR peptide of Human Bin1/Amphiphysin II

The plasmid of the BAR domain of human Bin1/Amphiphysin II was a kind gift of E.D. Laue (Cambridge). The histidine tagged recombinant protein was expressed in *E. coli* BL21(DE3) and purified as previously described (12). The N-BAR peptides were expressed as SUMO fusion protein and cleaved by the SUMO protease (23) or synthesized by solid-phase peptide synthesis. Further details are given in the supporting information (SI).

Liposomes and tubulation assay

Small unilamellar vesicles were prepared from total bovine brain lipids (Folch fraction 1, Sigma B1502) in 20 mM Hepes, 150 mM NaCl, pH 7.4 by extrusion (100 nm pore size) using a Liposofast extruder (Avestin, Ottawa, ON) as described (13). For tubulation assays N-BAR domain and different constructs (5 µM for N-BAR and BAR, 10 µM for N-BAR peptides) were mixed with brain lipid liposomes (0.2 mg/ml) for 30 min at room temperature and then processed for negative staining. For EM analysis, carbonized copper grids (Plano, Wetzlar, Germany) were pretreated for 1 min with bacitracin (0.1 mg/ml). After air drying, protein lipid mixture that had been diluted with 20 mM Hepes, 150 mM NaCl, pH 7.4 5-fold was applied for 3 min. Subsequently, grids were again air dried. Samples were negatively stained with 1 % (w/v) uranyl-acetate and visualized in a Zeiss EM 900 electron microscope operating at 80 kV.

FRET assay of membrane fusion

Membrane fusion was measured by fluorescence energy transfer using a JASCO FP6500 spectrometer. Two populations of liposomes composed of bovine brain lipids, one unlabeled and one labeled with 2 % each of *n*-[7-nitro-2-1,3-benzoaxadiazole-4-yl]-egg-phosphatidylethanolamine and *n*-[lissamine rhodamine B]-egg-phosphatidylethanolamine, were mixed at a 9:1 unlabeled/labeled ratio and 0.25 mg/ml total lipid in 20 mM Hepes, 150 mM NaCl, pH 7.4 at 25 °C in the presence of different concentrations of N-BAR and various constructs (concentration range: 1-15 µM for N-BAR and BAR, 5-50 µM for N-BAR peptide constructs). The excitation wavelength was 450 nm and the emission spectrum was recorded from 480-700 nm after several time points. 1 % Triton X-100 was added to obtain a value for donor fluorescence. Kinetics of membrane fusion were followed by fluorescence increase of the donor fluorescence at 530 nm after excitation at 450 nm.

Circular dichroism

Far-UV CD spectra of BAR domain and mutants were measured in the presence and absence of Folch liposomes in 20 mM Hepes, 150 mM NaCl, pH 7.4 at 15 °C with a JASCO J815A spectropolarimeter. 10 µM protein was incubated with 0.2 mg/ml Folch liposomes and degassed for 5 min before measurement. The N-terminal peptide was measured in the presence of different detergents (SDS, OG, DPC) and Folch liposomes. The signals from pure liposomes or micelles were subtracted from the sample spectra as blanks.

Sample preparation for NMR

N-BAR peptide was dissolved in 20 mM sodium-phosphate pH 7.4 (90 % H_2O/10 % ^2H_2O) containing either d_{38}-DPC or d_{25}-SDS micelles and 0.03 % NaN_3. The final ^{15}N-labeled N-BAR peptide samples contained 1 mM protein and 200 mM d_{38}-DPC or 150 mM d_{25}-SDS. A 1 mM ^{15}N-labeled N-BAR peptide sample without detergent was prepared as a reference.

Results and Discussion

The predicted N-terminal amphipathic helix-0 of the human Bin1/Amphiphysin II BAR domain comprises residues 1-33 (14,37). This part of the molecule is disordered in the absence of lipids and thereby unresolved in the crystal structure (12). The N-BAR peptide studied here (^{1}MAEMGSKGV^{10}TAGKIASNVQ^{20}KKLTRAQEKV^{30}LQKLY) contains an additional tyrosine at the C-terminus of helix-0 for spectroscopic reasons. DPC and SDS micelles were chosen, because they have been successfully employed for other peptide and protein structures determinations by NMR spectroscopy (38-41).

Structure induction upon membrane binding

Far-UV CD spectra of the full length N-BAR domain revealed an increased helicity after adding brain lipid liposomes (solid lines in Fig. 1a). The helical content of a deletion mutant, lacking the first 31 residues (BAR) did not change in the presence of lipids. The isolated N-BAR peptide is unstructured in aqueous solution (solid black line in Fig. 1b). As in the full length protein, the ellipticity minima at 208 and 222 nm indicate that the peptide takes on helical structure when bound to liposomes or micelles. The CD spectra of the N-BAR peptide in brain lipid liposomes, DPC or SDS micelles are virtually identical, indicating a similar secondary structure under these conditions (Fig. 1b). In 60 % TFE the helical content increased further, indicating that not all residues of the N-BAR peptide are in a helical conformation in the presence of detergents or lipids. A helical wheel projection (Fig. 3d) of the peptide highlights its amphipathic character (i.e., hydrophobic and charged/polar residues are located opposite of each other). Conspicuous is the high number of lysine residues, implying that binding and structure induction of the N-BAR peptide is not just driven by hydrophobic but also electrostatic interactions. This is further supported by the observation that non-ionic OG micelles as membrane mimic do not lead to structuring of the peptide as judged by the far-UV CD spectrum (dashed light grey in Fig. 1b).

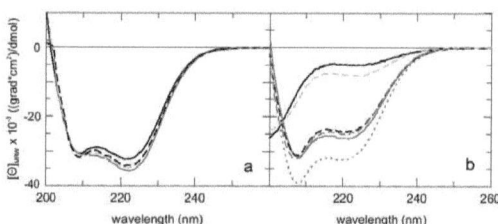

Figure 1. Far-UV CD spectra of the BAR domain and the N-terminal N-BAR peptide in various solvent environments. (a) CD spectra of the N-BAR (residues 1-241) (solid line) and the BAR (residues 32-241) domain (dashed line) of human amphiphysin II in the presence and absence of brain lipid liposomes (grey and black respectively). Structure induction upon binding to liposomes is only seen for the N-BAR domain (solid grey line), indicated by a significant signal decrease at 222 nm. (b) The N-BAR peptide is unstructured in solution (solid black line). In the presence of liposomes (dashed dark grey line), SDS (solid grey line) or DPC (black dashed line) micelles, the peptide becomes structured. In the presence of OG (dashed light grey line) micelles however, no structure induction is observed. A CD spectrum recorded in 60 % TFE (dotted line) shows the highest helical content.

NMR structure and dynamics of N-BAR peptide

Binding of the N-BAR peptide to micelles results in a deviation of the backbone and more obviously the side chain resonances in the ^{15}N-HSQC spectrum (Fig. 2b) from the random coil chemical shifts dominating the spectrum in aqueous solution (Fig. 2a). This confirms the interaction of N-BAR with the micellar environment and induction of defined secondary structure in SDS or DPC observed by far UV-CD.

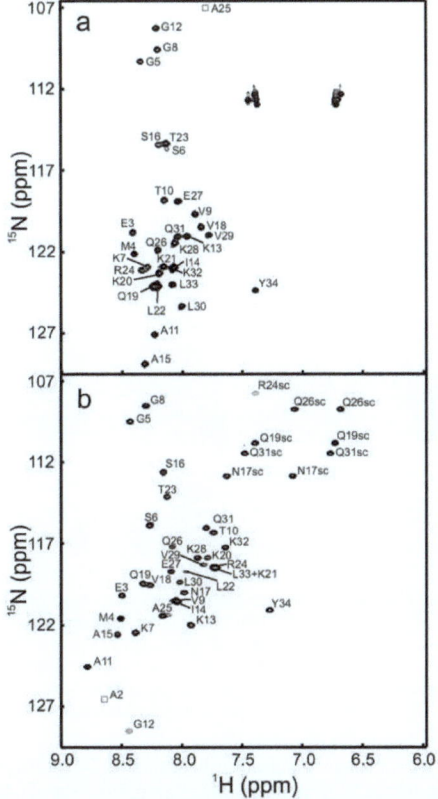

Figure 2. 2D ^{1}H-^{15}N HSQC spectra of the N-BAR peptide (a) in aqueous solution, and (b) bound to DPC micelles. The assigned cross-peaks of the backbone amides are labeled using the one-letter amino acid code and the sequence position. Boxes indicate resonance signals, which show cross peak intensities below the plotted contour level. The respective spectrum of N-BAR peptide in SDS micelles is shown in the supporting information (SI) Fig. 12.

All backbone and side chain resonances were assigned as described in Material and Methods. More than 500 NOE distance constraints were derived from 2D-NOESY and ^{15}N-HSQC-NOESY spectra in the presence of either SDS or DPC micelles. By using all experimentally determined constraints (NOEs, dihedral angles derived from J couplings and chemical shifts), ensembles of structures of the N-BAR peptide in SDS (Fig. 3a) and DPC (Fig. 3b) micelles were calculated (structural statistics are given in the SI Table 2). Residues 8-34 in SDS and

10-34 in DPC micelles are well ordered with a heavy atom r.m.s. deviation below 1.1 Å. This α-helical content is consistent with the CD data (Fig. 1). Hence, the structured part of the N-BAR peptide is an amphipathic helix with the negatively charged side chains on the convex side and the hydrophobic side chains on the concave side (Fig. 3c).

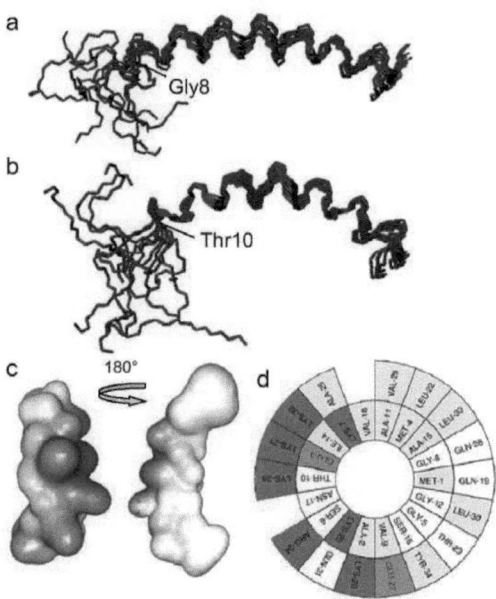

Figure 3. Structure ensembles of the N-BAR peptide backbone bound to detergent micelles at 25 °C: 10 lowest energy structures in (a) SDS micelles and (b) DPC micelles. (c) Electrostatic surface potential representation of the N-BAR peptide in DPC micelles. Negative potentials are shown in red and positive potentials in blue. (d) Helical wheel diagram for the N-BAR peptide. The amino acid sequence is plotted clockwise. Hydrophobic residues are shown in grey boxes and positively and negatively charged residues in blue and red, respectively.

The disordering of the N-terminus results from fast and large amplitude local dynamics confirmed by ^{15}N-heteronuclear NOE (hNOE) measurements (Fig. 4). hNOE values above 0.5 are typical for structural elements in peptides and proteins that are relatively rigid on a nanoseconds-to-picoseconds time scale. For the N-BAR peptide in SDS and DPC micelles, the hNOE gradually decreases from T10 towards the N-terminus and is even negative for the first residues. Therefore, the dynamic data agree well with the loss of NOE constraints in the highly flexible and disordered conformation at the N-terminus. In comparison, all hNOE values of the N-BAR peptide in aqueous solution (Fig. 4c) are close to zero or negative. Together with a lack of medium range NOEs over the entire sequence, a random coil conformation in the absence of detergent and lipids can be concluded. The hNOE of an extended N-BAR peptide in DPC micelles with 44 amino acids drops after K35 towards the C-terminus indicating that the amphipathic helix ends at position 35 and following residues form the linker to helix-1 of the BAR domain.

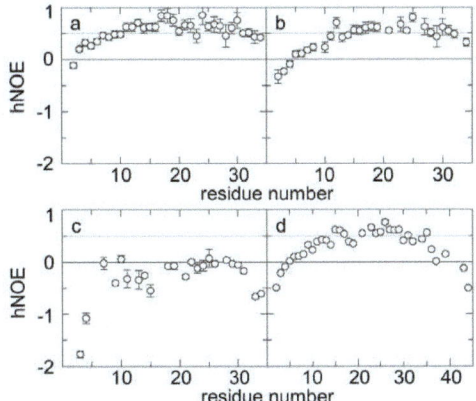

Figure 4. ^{1}H-^{15}N heteronuclear NOEs of the N-BAR peptide in (a) SDS, (b) DPC and (c) aqueous solution. (d) hNOE values of the extended N-BAR peptide (1-44 residues).

Fast ms-amide proton exchange

To determine which regions of the N-BAR peptide are buried in the detergent micelle, fast (millisecond) amide-protein exchange was measured for each residue by NMR (24); examples of exchange curves are depicted in Fig. 5. This approach is straightforward compared to the use of spin labels, because the NMR sample for structure determination can be used without further modifications. N-terminal, polar and charged residues showed a pronounced signal change during the experiment (color coded in red) indicating fast exchanging amide protons because of an increased solvent accessibility and a dynamic open and closing of the corresponding hydrogen bonds on the ms-s timescale. Amide protons of hydrophobic residues however, did not exchange at all (color coded in blue), because they are buried in the micelle and therefore shielded from the solvent. Furthermore, residues 1-11, the region with high local fluctuations derived from dynamic data (Fig. 4b,c), show low protection against exchange of the amide protons.

Figure 5. NMR experiment to detect fast exchanging amide protons (MEXICO) of the N-BAR peptide bound to SDS and DPC micelles. Fast amide proton exchange was followed on a residue by residue level. (a) Exchange curves in SDS micelles are shown for T23 (closed red symbols), S16 (open red symbols), L33 (closed blue symbols) and V29 (open blue symbols). Fast exchanging amides are colored in red. Amide protons, which did not exchange within the timescale of the experiment (below dashed line, see also SI Fig. 13) are colored in blue. Exchange curves for residues in grey could not been evaluated due to signal overlap or low signal intensity. This color code was assigned to ribbon representation the lowest energy NMR structure of the N-BAR peptide in (b) SDS and (c) DPC micelles.

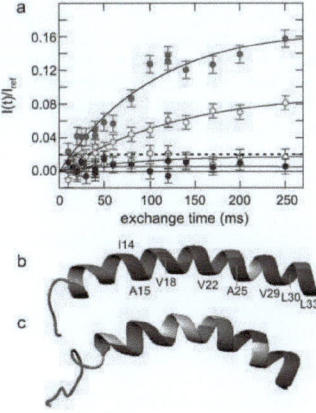

MD simulation of the N-BAR peptide

As a complement to the preceding experimental results, molecular dynamics simulations were carried out on the N-BAR peptide in micelles with different surfactants, micelle sizes, and initial configurations to investigate helical stability, peptide orientation, and depth of insertion in the micelle environment. Fig. 6 shows snapshots at the beginning (left) and end (right) of three of the five 60 ns simulations of peptide/micelle complex simulations (Table 1).

Table 1. Initial conditions for the N-BAR peptide/micelle simulations.

Name	Micelles		Initial position of the peptide	
	Type	No. of molecules	Surrounding environment	Direction of hydrophobic core
SDS1	SDS	75	Outside micelle	Micelle
SDS2	SDS	75	Outside micelle	Water
SDS3	SDS	75	Inside micelle	-
SDS4	SDS	40	Inside micelle	-
DPC	DPC	65	Inside micelle	-

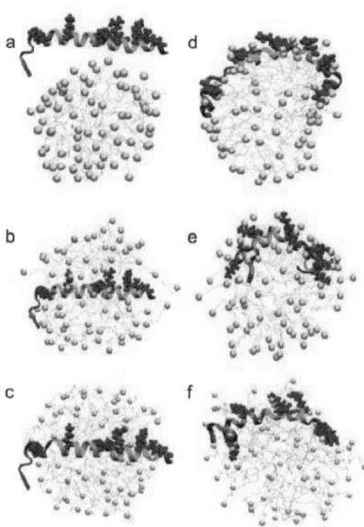

Figure 6. Snapshots at the beginning (left) and end (right) of 60 ns simulations of the peptide/micelle systems denoted (a,d) SDS1, (b,e) SDS3, and (c,f) DPC (Table 1). α-Helical regions of the peptides are presented in green, non-helical in gray, and positively charged side chains in blue. The negatively charged sulfur of SDS and phosphorous of DPC are yellow, and the acyl chains are light blue. The N-BAR peptide is initially positioned inside or outside a micelle. Water and ions are omitted for clarity. The images were created using VMD (42).

In each case, the peptide migrated to the surface of the micelle. The micelle remained spherical and the peptide curved. In pure water the N-BAR peptide partially unfolds in the first ns, and loses almost all helicity by 13 ns (see SI Fig. 9). Hence, the micelle environment stabilizes the helix in most cases, leading to a range of α-helicities of 40-50 % (SDS1: 48 %, SDS3: 50 %, DPC: 40 %) for residues 12-30 during the last simulated 20 ns. The helical instability of the N-terminal residues agrees with the amide proton exchange and the NMR relaxation data (Fig. 4 and Fig. 5). The SDS4 simulation (the smaller micelle; see SI Fig. 10) yielded a slightly lower helical content, 30 %. It is possible, that the higher curvature imposed on the peptide for binding led to this instability. During the SDS2 run (see SI Fig. 10), the N-BAR peptide bound to the micelle with little (6 %) α-helicity. This can be attributed to the initial condition, where the cationic residues of the peptide were oriented towards the micelle. When the peptide interacts with the micelle, cationic and hydrophobic residues respectively have favorable interactions with SDS head groups and tails, which may lead to a flip of the peptide and instability of the helical structure. It is possible that the peptide in SDS2 would refold to the helical form in much longer simulations, but this is outside the scope of the present study. These results indicate that MD simulations can represent the experimentally measured stability of the peptide, although the final configuration is partially determined by initial configuration and micelle size. Further analyses are based on SDS1, SDS3 and DPC.

Orientation of the peptide in micelles
Experimental results from fast proton exchange experiments of backbone amides are compared in Fig. 7a with the solvent accessible surface area (SASA) calculated for SDS3. Between A11 and Y34, the SASA correlates well with the hydrophobic residues (squared symbols) and residues with low exchange rates (black symbols) buried in the micelle, whereas charged residues face towards water. This correlation is less pronounced for the DPC micelle (SI Fig. 11). To investigate factors controlling the depth of penetration of the N-BAR peptide, radial distribution functions (g(r)) between charged residues of the peptide and lipid head groups were calculated for SDS3 and DPC. The integral of g(r) over a particular interval is proportional to the coordination number in that interval. Fig. 7b,c shows that g(r) of the lipid head groups around cationic residues of the peptide is substantially higher than for anionic residues. Moreover, in the DPC micelles, g(r) of the phosphate head groups around cationic residues of the peptide have higher values compared to the choline head groups around both anionic and cationic residues of the peptide. These results imply that cationic residues of the peptide strongly interact with anionic lipid head groups, and that the N-BAR peptide embeds deeper in the DPC micelle compared to SDS.

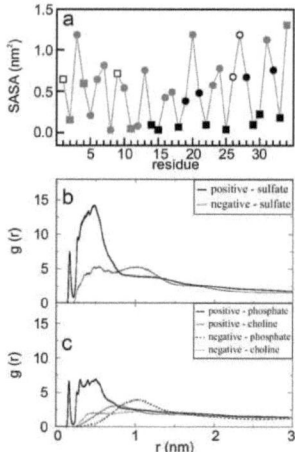

Figure 7. (a) Solvent accessible surface areas per residue of the N-BAR peptide in SDS micelles (SDS3 simulation). Squared symbols represent hydrophobic residues shown in Fig. 3d. Black symbols correspond to residues not exposed to water according to the amide proton exchange experiments, grey denotes fast exchanging protons, and white is for missing experimental data. Radial distribution functions g(r) for charged residues of the peptide with respect to SDS head groups in SDS3 (b) and DPC head groups in DPC (c) averaged over the last 20 ns.

Tubulation of liposomes
The ability of BAR domains to generate membrane curvature has been shown *in vitro* (13,17,20,21,43) by the formation of narrow tubes of liposomes. To investigate the significance of helix-0 in this process, we determined changes in the shape of liposomes in the presence of different BAR domain constructs by electron microscopy. The human amphiphysin N-BAR domain is able to constrict liposomes into tubules, but leads to vesiculation at higher protein concentrations (Fig. 8b) (13). A deletion mutant, lacking the N-terminal amphipathic helix (BAR) had a much smaller influence on the liposome morphology (Fig. 8c). In addition to extensive vesiculation, tube formation was observed for the N-BAR peptide (Fig. 8d,e) and the extended 1-44 residue N-BAR peptide (Fig. 8f). Fluorescence energy transfer (FRET) based membrane fusion assays yield a more quantitative measure of the membrane fusion properties of the different BAR domain constructs. Liposomes were prepared with fluorescence labeled lipids and subsequently mixed with unlabeled liposomes; fusion of labeled with unlabeled liposomes can be followed by the quench of the FRET-signal concomitant with an increase of the donor fluorescence at 530 nm. Fig. 8g,h shows membrane fusion for the N-BAR domain and the N-BAR peptides, but not for the BAR domain lacking helix-0.

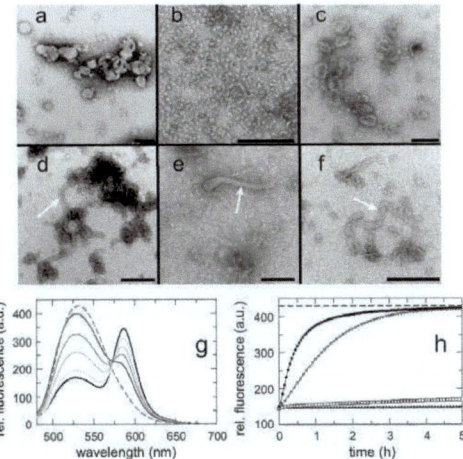

Figure 8. Electron micrographs of liposome tubulation by (b) human amphiphysin N-BAR, (c) BAR and (d,e) the N-BAR peptide (length scale, black bar, 200 nm). Untreated liposomes are shown in (a) and in the presence of an extended N-BAR peptide (residues 1-44) in (f). Emission spectra (g) from mixed liposomes in the absence (black curve) and presence of the N-BAR domain at various time points (8 min, 28 min, 62 min, 225 min, light to dark grey) and 1 % Triton X-100 (for total donor fluorescence, dashed line). (h) Time dependent increase of donor fluorescence at 530 nm upon membrane fusion in the presence of N-BAR (●), BAR (□) and the N-BAR peptide (▽). Fluorescence change caused by spontaneous liposome fusion is negligible (▲) and at maximum in 1 % Triton X-100 (dashed line).

Conclusions

The present study combines experimental and theoretical techniques to obtain detailed insights into the structural properties of helix-0 of human Bin1/Amphiphysin II BAR domain. Since structural information for the interaction of proteins and lipids requires the introduction of mutations and spin labels (20), the N-BAR peptide was investigated when bound to detergent micelles and during tubulation of liposomes. Experiment and simulation confirmed the predicted random coil to helix transition upon micelle binding but revealed an unstructured and solvent exposed N-terminal region. Binding and amphipathic structure induction is mediated by hydrophobic as well as electrostatic interactions. The MD simulations complete the structural view of the N-BAR peptide in SDS and DPC micelle derived from the NMR structure calculations in vacuum. The robustness of the simulations is reflected by comparable final conformations of the N-BAR peptide starting from different conditions and that the peptide curvature, which was maintained during the simulation, adopts to the micelle. A slightly increased curvature of the N-BAR peptide in DPC micelles compared to SDS found in the calculated NMR structures (Fig. 3a,b) is a result of several side chain-side chain and backbone-side chain NOEs, which were unambiguously identified. This curvature might result from the deeper embedded peptide caused by the polar interactions with the zwitterionic DPC head groups, if we assume the same spherical size of both micelles types.

Recent experiments based on fluorescence measurements imply an antiparallel dimer formation of the helix-0 of the BRAP/Bin2 BAR domain (breast-cancer-associated protein) domain when bound to liposomes (14). In the present study, no long-range NOEs between N- and C-terminal residues of the N-BAR peptide were observed in the NOESY spectra, which rules out dimer formation. Nevertheless oligomerization of the helix-0 under different conditions can not be excluded since the structure was determined in detergent micelles with a high detergent/peptide ratio.

Recently, a point mutation (K35N) in the helix-0 of the human Bin1/Amphiphysin II N-BAR domain was found in autosomal recessive centronuclear myopathy (11), which showed different tubulation properties in *ex vivo* membrane assays. In isolation at least, extended N-BAR peptides comprising 1-44 residues with K35 and N35 could not be distinguished from the 1-34 peptide in their here presented biophysical properties (CD and NMR spectra, tubulation). The tubulation experiments revealed the importance of helix-0 for changing membrane morphology of liposomes by itself or when present in the BAR domain.

In summary, the present results highlight the importance of formation of amphipathic helix-0 to increase the affinity of the N-BAR domain to lipid bilayers. This supports the currently discussed models about the curvature-generating mechanism, which are based on this interaction. While the present findings do not definitively rule out the spontaneous curvature and bilayer-coupling mechanics for curvature generation, they favor the scaffold mechanism, because the N-BAR peptide adopts during the MD simulations to the micelle environment rather than disordering it. Therefore we expect curving-generation not before the main interaction with the entire N-BAR domain.

Acknowledgement

We thank Paul Rösch for NMR spectrometer time at 600, 700 and 800 MHz, Gerd Hause and Rolf Sachs for electron microscopy, Andreas Kerth for help with liposome preparation and Alfred Blume for helpful discussions. This research was supported in part by a grant from the Deutsche Forschungsgemeinschaft (Ba 1821/3-1 and GRK 1026), the excellence initiative of the state Sachsen-Anhalt, the Intramural Research Program of the NIH, National Heart, Lung and Blood Institute, and utilized the high-performance computational capabilities of the CIT Biowulf/LoBoS3 cluster at NIH.

References

1. David, C., P.S. McPherson, O. Mundigl, and P. de Camilli. 1996. A role of amphiphysin in synaptic vesicle endocytosis suggested by its binding to dynamin in nerve terminals. Proc. Natl. Acad. Sci. U S A 93:331-335.
2. Shupliakov, O., P. Low, D. Grabs, H. Gad, H. Chen, C. David, K. Takei, P. De Camilli, and L. Brodin. 1997. Synaptic vesicle endocytosis impaired by disruption of dynamin-SH3 domain interactions. Science 276:259-263.
3. Ren, G., P. Vajjhala, J.S. Lee, B. Winsor, and A.L. Munn. 2006. The BAR domain proteins: molding membranes in fission, fusion, and phagy. Microbiol. Mol. Biol. Rev. 70:37-120.
4. Di Paolo, G., S. Sankaranarayanan, M.R. Wenk, L. Daniell, E. Perucco, B.J. Caldarone, R. Flavell, M.R. Picciotto, T.A. Ryan, O. Cremona, and P. De Camilli. 2002. Decreased synaptic vesicle recycling efficiency and cognitive deficits in amphiphysin 1 knockout mice. Neuron 33:789-804.
5. Zhang, B., and A.C. Zelhof. 2002. Amphiphysins: raising the BAR for synaptic vesicle recycling and membrane dynamics. Bin-Amphiphysin-Rvsp. Traffic 3:452-460.
6. Lee, E., M. Marcucci, L. Daniell, M. Pypaert, O.A. Weisz, G.C. Ochoa, K. Farsad, M.R. Wenk, and P. De Camilli. 2002. Amphiphysin 2 (Bin1) and T-tubule biogenesis in muscle. Science 297:1193-1196.
7. McMahon, H.T., and J.L. Gallop. 2005. Membrane curvature and mechanisms of dynamic cell membrane remodelling. Nature 438:590-596.
8. Dawson, J.C., J.A. Legg, and L.M. Machesky. 2006. Bar domain proteins: a role in tubulation, scission and actin assembly in clathrin-mediated endocytosis. Trends Cell Biol. 16:493-498.
9. Shimada, A., H. Niwa, K. Tsujita, S. Suetsugu, K. Nitta, K. Hanawa-Suetsugu, R. Akasaka, Y. Nishino, M. Toyama, L. Chen, Z.J. Liu, B.C. Wang, M. Yamamoto, M. Terada, A. Miyazawa, A. Tanaka, S. Sugano, M. Shirouzu, K. Nagayama, T. Takenawa, and S. Yokoyama. 2007. Curved EFC/F-BAR-domain dimers are joined end to end into a filament for membrane invagination in endocytosis. Cell 129:761-772.
10. Henne, W.M., H.M. Kent, M.G. Ford, B.G. Hegde, O. Daumke, P.J. Butler, R. Mittal, R. Langen, P.R. Evans, and H.T. McMahon. 2007. Structure and analysis of FCHo2 F-BAR domain: a dimerizing and membrane recruitment module that effects membrane curvature. Structure 15:839-852.
11. Nicot, A.S., A. Toussaint, V. Tosch, C. Kretz, C. Wallgren-Pettersson, E. Iwarsson, H. Kingston, J.M. Garnier, V. Biancalana, A. Oldfors, J.L. Mandel, and J. Laporte. 2007. Mutations in amphiphysin 2 (BIN1) disrupt interaction with dynamin 2 and cause autosomal recessive centronuclear myopathy. Nat. Genet. 39:1134-1139.
12. Casal, E., L. Federici, W. Zhang, J. Fernandez-Recio, E.M. Priego, R.N. Miguel, J.B. DuHadaway, G.C. Prendergast, B.F. Luisi, and E.D. Laue. 2006. The crystal structure of the BAR domain from human Bin1/amphiphysin II and its implications for molecular recognition. Biochemistry 45:12917-12928.
13. Peter, B.J., H.M. Kent, I.G. Mills, Y. Vallis, P.J. Butler, P.R. Evans, and H.T. McMahon. 2004. BAR domains as sensors of membrane curvature: the amphiphysin BAR structure. Science 303:495-499.
14. Fernandes, F.M., L.M. Loura, F.J. Chichon, J.L. Carrascosa, A. Fedorov, and M. Prieto. 2008. Role of Helix-0 of the N-BAR domain in membrane curvature generation. Biophys J:in press.
15. Takei, K., V.I. Slepnev, V. Haucke, and P. De Camilli. 1999. Functional partnership between amphiphysin and dynamin in clathrin-mediated endocytosis. Nat. Cell Biol. 1:33-39.
16. Blood, P.D., and G.A. Voth. 2006. Direct observation of Bin/amphiphysin/Rvs (BAR) domain-induced membrane curvature by means of molecular dynamics simulations. Proc. Natl. Acad. Sci. U S A 103:15068-15072.
17. Farsad, K., N. Ringstad, K. Takei, S.R. Floyd, K. Rose, and P. De Camilli. 2001. Generation of high curvature membranes mediated by direct endophilin bilayer interactions. J. Cell Biol. 155:193-200.
18. Ford, M.G., I.G. Mills, B.J. Peter, Y. Vallis, G.J. Praefcke, P.R. Evans, and H.T. McMahon. 2002. Curvature of clathrin-coated pits driven by epsin. Nature 419:361-366.
19. Lee, M.C., L. Orci, S. Hamamoto, E. Futai, M. Ravazzola, and R. Schekman. 2005. Sar1p N-terminal helix initiates membrane curvature and completes the fission of a COPII vesicle. Cell 122:605-617.
20. Gallop, J.L., C.C. Jao, H.M. Kent, P.J. Butler, P.R. Evans, R. Langen, and H.T. McMahon. 2006. Mechanism of endophilin N-BAR domain-mediated membrane curvature. EMBO J. 25:2898-2910.
21. Masuda, M., S. Takeda, M. Sone, T. Ohki, H. Mori, Y. Kamioka, and N. Mochizuki. 2006. Endophilin BAR domain drives membrane curvature by two newly identified structure-based mechanisms. EMBO J. 25:2889-2897.
22. Zimmerberg, J., and M.M. Kozlov. 2006. How proteins produce cellular membrane curvature. Nat. Rev. Mol. Cell Biol. 7:9-19.
23. Bosse-Doenecke, E., U. Weininger, M. Gopalswamy, J. Balbach, S. Möller Knudsen, and R. Rudolph. 2008. High yield production of recombinant native and modified peptides exemplified by ligands for G-protein coupled receptors. Protein Expr. Purif. 58:114-121.
24. Koide, S., W. Jahnke, and P.E. Wright. 1995. Measurement of intrinsic exchange rates of amide protons in a 15N-labeled peptide. J. Biomol. NMR 6:306-312.

25. Delaglio, F., S. Grzesiek, G.W. Vuister, G. Zhu, J. Pfeifer, and A. Bax. 1995. NMRPipe: a multidimensional spectral processing system based on UNIX pipes. J. Biomol. NMR 6:277-293.
26. Johnson, B.A. 2004. Using NMRView to visualize and analyze the NMR spectra of macromolecules. Methods Mol. Biol. 278:313-352.
27. Linge, J.P., M. Habeck, W. Rieping, and M. Nilges. 2003. ARIA: automated NOE assignment and NMR structure calculation. Bioinformatics 19:315-316.
28. Laskowski, R.A., J.A. Rullmannn, M.W. MacArthur, R. Kaptein, and J.M. Thornton. 1996. AQUA and PROCHECK-NMR: programs for checking the quality of protein structures solved by NMR. J. Biomol. NMR 8:477-486.
29. Lindahl, E., B. Hess, and D. van der Spoel. 2001. GROMACS 3.0: A package for molecular simulation and trajectory analysis J. Mol. Mod. 7:306-317.
30. van der Spoel, D., E. Lindahl, B. Hess, G. Groenhof, A.E. Mark, and H.J. Berendsen. 2005. GROMACS: fast, flexible, and free. J. Comput. Chem. 26:1701-1718.
31. Schüttelkopf, A.W., and D.M. van Aalten. 2004. PRODRG: a tool for high-throughput crystallography of protein-ligand complexes. Acta Crystallogr. D Biol. Crystallogr. 60:1355-1363.
32. MacKerell Jr, A.D. 1995. Molecular dynamics simulation analysis of a sodium dodecyl sulfate micelle in aqueous solution: decresed fluidity of the micelle hydrocarbon interior. J. Phys. Chem. 99:1846-1855.
33. Jorgensen, W.L., and J. Tirado-Rives. 1988. The OPLS potential functions for proteins. Energy minimization for crystals of cyclic peptides and crambin. J. Am. Chem. Soc. 110:1657-1666.
34. Bales, B.L., and M. Almgren. 1995. Fluorescence quenching of pyrene by copper (II) in sodium dodecyl sulfate micelles. Effect of micelle size as controlled by surfactant concentration. J. Phys. Chem. 99:15153-15162.
35. Berendsen, H.J.C., J.P.M. Postma, W.F. van Gunsteren, A. Dinola, and J.R. Haak. 1984. Molecular dynamics with coupling to an external bath. J. Chem. Phys. 81:3684-3690.
36. Kabsch, W., and C. Sander. 1983. Dictionary of protein secondary structure: pattern recognition of hydrogen-bonded and geometrical features. Biopolymers 22:2577-2637.
37. Gallop, J.L., and H.T. McMahon. 2005. BAR domains and membrane curvature: bringing your curves to the BAR. Biochem. Soc. Symp.:223-231.
38. Han, X., J.H. Bushweller, D.S. Cafiso, and L.K. Tamm. 2001. Membrane structure and fusion-triggering conformational change of the fusion domain from influenza hemagglutinin. Nat. Struct. Biol. 8:715-720.
39. Liang, B., and L.K. Tamm. 2007. Structure of outer membrane protein G by solution NMR spectroscopy. Proc. Natl. Acad. Sci. U S A 104:16140-16145.
40. Kessler, H., D.F. Mierke, J. Saulitis, S. Seip, S. Steuernagel, T. Wein, and M. Will. 1992. The structure of Ro 09-0198 in different environments. Biopolymers 32:427-433.
41. Koppitz, M., B. Mathä, and H. Kessler. 1999. Structure investigation of amphiphilic cyclopeptides in isotropic and anisotropic environments-A model study simulating peptide-membrane interactions. J. Pept. Sci. 5:507-518.
42. Humphrey, W., A. Dalke, and K. Schulten. 1996. VMD: visual molecular dynamics. J. Mol. Graph. 14:33-38.
43. Richnau, N., A. Fransson, K. Farsad, and P. Aspenstrom. 2004. RICH-1 has a BIN/Amphiphysin/Rvsp domain responsible for binding to membrane lipids and tubulation of liposomes. Biochem. Biophys. Res. Commun. 320:1034-1042.
44. Koglin, N., M. Lang, R. Rennert, and A.G. Beck-Sickinger. 2003. Facile and selective nanoscale labeling of peptides in solution by using photolabile protecting groups. J Med Chem 46:4369-4372.

Subproject E

Supporting Information (SI)

Expression and purification of the N-BAR and BAR Domain and the N-BAR peptide of Human Bin1/Amphiphysin II. The plasmid of the BAR domain of human Bin1/Amphiphysin II was a kind gift of E.D. Laue (Cambridge). The histidine tagged recombinant protein was expressed in *E. coli* BL21(DE3) and purified as previously described (12). The N-BAR peptide was synthesized by automated solid-phase peptide synthesis using Fmoc/^{t}Bu-strategy on a Syro Multipeptide Robot System. Wang-Resin was used for anchoring and Fmoc-protected amino acids were coupled twice with 10-fold excess each. For activation diisopropylcarbodiimide and 1-hydroxybenzotrialzole were used. Cleavage was performed with trifluoroacetic acid and a scavenger mixture of thiocresole/thioanisol (90:5:5) within 3 h, followed by precipitation from diethyl ether and repetitive washing steps (44). After lyophilization from water/tert. butanol, the peptide was purified by preparative HPLC and characterized by MALDI-TOF mass spectrometry and analytical HPLC. Purity was > 98%, elution occurred at 39.2 % acetonitrile containing 0.08% trifluoroacetic acid, experimentally found mass $[M+H]^+$ 3544.1 Da corresponded with the theoretically calculated mass of $[M+H]^+ = 3544.29$ Da.

To facilitate isotope labeling, the N-BAR peptide was expressed as SUMO fusion protein and cleaved by the SUMO protease (23). A deletion mutant lacking the first 31 residues (BAR) was cloned into the bacterial expression vector pET14b, expressed, and purified as the wild type protein. The N-terminal residues 1-33 and 1-43 were amplified from the wild type plasmid with flanking primers and cloned into a modified petSUMO vector (23) using BsaI and BamH1 restriction sites. An additional tyrosine residue at position 34 and 44 was introduced for concentration determination. The gene sequence was confirmed by automated DNA sequencing. The fusion protein was expressed in *E. coli* BL21 (DE3) and purified from soluble material. Cells were resuspended in IMAC binding buffer (50 mM sodium-phosphate, 300 mM NaCl, 20 mM imidazole, pH 8.0) and lysed by sonication. Protein was eluted from the IMAC-column by step elution with 250 mM imidazole and dialyzed against SUMO protease cleavage buffer. The fusion protein was cleaved with specific SUMO protease at 4 °C overnight (0.1 mg SUMO protease per 10 mg fusion protein). Cleaved material was subjected to a second IMAC step and the flow through was further purified by RP-HPLC on a C18 column (SP 250/10 Nucleosil 500-5 C18 PPN, Macherey-Nagel). Peak fractions were collected and the identity confirmed by electro spray mass spectrometry. The pooled fractions were lyophilized and stored at -20 °C. CD spectra of synthesized and recombinant N-BAR peptide were identical. Isotope labeled ^{15}N-NMR-samples were produced using M9 minimal media made up with ^{15}NH$_4$Cl as nitrogen source and supplemented with vitamin mix.

References:

1. Casal, E., L. Federici, W. Zhang, J. Fernandez-Recio, E.M. Priego, R.N. Miguel, J.B. DuHadaway, G.C. Prendergast, B.F. Luisi, and E.D. Laue. 2006. The crystal structure of the BAR domain from human Bin1/amphiphysin II and its implications for molecular recognition. Biochemistry 45:12917-12928.
2. Koglin, N., M. Lang, R. Rennert, and A.G. Beck-Sickinger. 2003. Facile and selective nanoscale labeling of peptides in solution by using photolabile protecting groups. J. Med. Chem. 46:4369-4372.
3. Bosse-Doenecke, E., U. Weininger, M. Gopalswamy, J. Balbach, S. Möller Knudsen, and R. Rudolph. 2008. High yield production of recombinant native and modified peptides exemplified by ligands for G-protein coupled receptors. Protein Expr. Purif. 58:114-121.

SI Table 2. Structural statistics of the N-BAR peptide in SDS and DPC micelles at 25 °C

	SDS	DPC
Experimental NMR constraints		
Total constraints	758	594
NOE distance constraints	684	550
J-coupling constraints	26	-
Dihedral constraints	48	44
NOE constraint violations		
NOE > 0.5 Å	0.10 ± 0.30	0.00 ± 0.00
NOE > 0.3 Å	0.60 ± 0.49	0.00 ± 0.00
NOE > 0.1 Å	4.20 ± 1.83	4.90 ± 1.58
Molecular dynamics statistics		
Average energy (kJ/mol)	E_{tot} 44.6 ± 5.6	E_{tot} 37.7 ± 1.5
	E_{bond} 1.7 ± 0.3	E_{bond} 2.1 ± 0.2
	E_{angle} 20.9 ± 1.6	E_{angle} 20.7 ± 0.6
	$E_{improper}$ 2.7 ± 0.4	$E_{improper}$ 2.7 ± 0.2
	E_{NOE} 13.6 ± 5.2	E_{NOE} 11.8 ± 1.4
	E_{cdih} 1.1 ± 0.2	E_{cdih} 0.4 ± 0.1
	$E_{couplings}$ 4.7 ± 0.9	-
r.m.s.d. from ideal distance (Å)	Bonds 0.00181 ± 0.00018	Bonds 0.00200 ± 0.00008
	NOE 0.0197 ± 0.0056	NOE 0.0207 ± 0.0016
r.m.s.d. from ideal angles (degree)	Angles 0.372 ± 0.016	Angles 0.371 ± 0.006
	Improper Angles 0.265 ± 0.020	Improper Angles 0.267 ± 0.010
Atomic r.m.s. deviation from mean structure (Å)		
Backbone atoms of all residues	3.59 ± 1.29	3.77 ± 0.78
Heavy atoms of all residues	3.73 ± 1.14	3.99 ± 0.75
Backbone atoms of residues 6-34	0.55 ± 0.19	0.21 ± 0.10
Heavy atoms of residues 6-34	1.19 ± 0.19	0.75 ± 0.10
Ramachandran statistics analyzed using PROCHECK	Residues 6-34 most allowed 91.6 % allowed 8.4 %	Residues 8-34 most allowed 98.1 % allowed 1.9 %

SI Figure 9. Secondary structure profiles for the N-BAR peptide of simulations SDS1, SDS2, SDS3, SDS4, DPC, and N-BAR (aqueous solution)

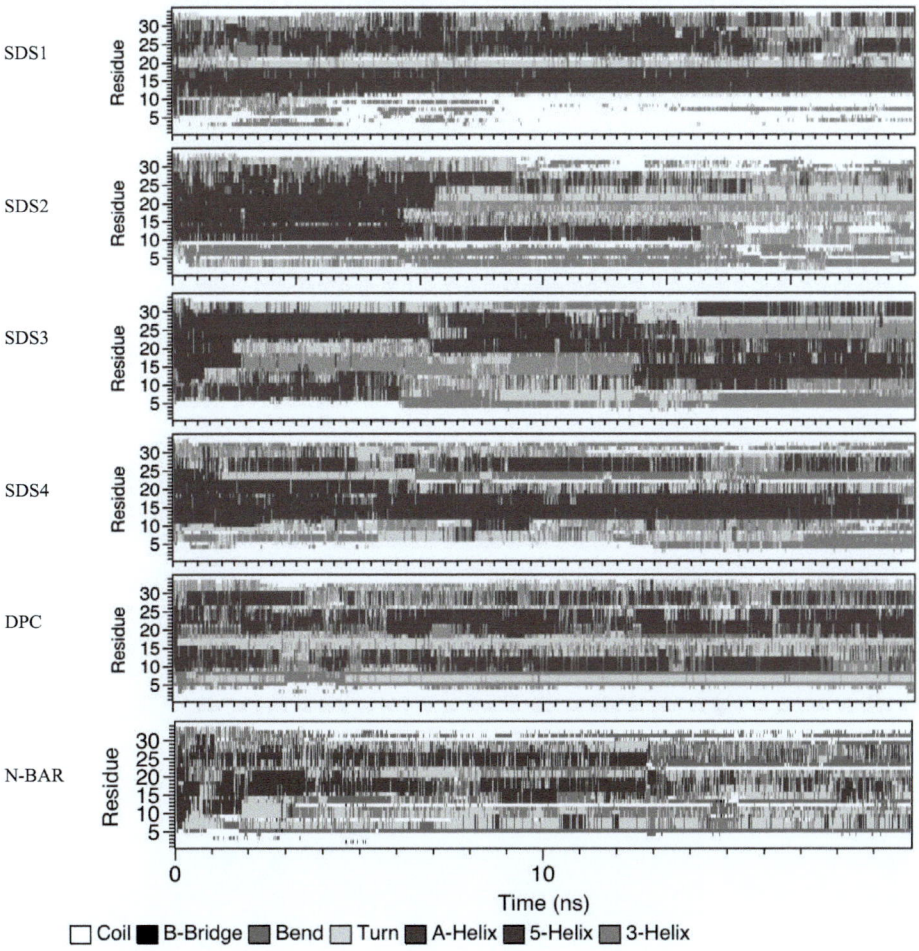

SI Figure 10. Snapshots at the beginning (left) and end (right) of 60 ns simulations of the peptide/micelle systems denoted SDS2 and SDS4 (see Table 1). α-Helical regions of the peptides are presented in green, with positively charged residues in blue. The negatively charged sulfur of SDS and phosphorous of DPC are yellow, and the acyl chains are light blue. The peptide is initially positioned inside or outside a micelle. Water and ions are omitted for clarity. The images were created using VMD (9).

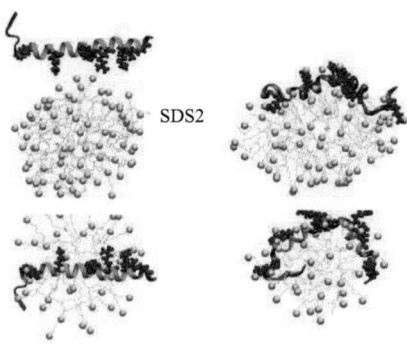

SI Figure 11. Solvent accessible surface areas per residue of the N-BAR peptide in DPC micelles (DPC simulation). Squared symbols represent hydrophobic residues according to Fig. 1c. Black symbols correspond to residues not exposed to water according to the amide proton exchange experiments, whereas gray codes for fast exchanging protons and white for missing experimental data.

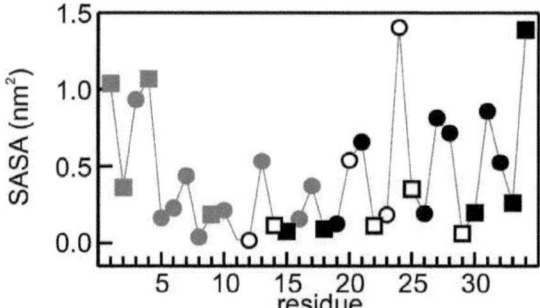

SI Figure 12. 2D ^{1}H-^{15}N HSQC spectra of the N-BAR peptide bound to SDS micelles. The assigned cross-peaks of the backbone amides are labeled using the one-letter amino acid code and the sequence position. Grey cross peaks are negative because of spectral folding in the ^{15}N dimension.

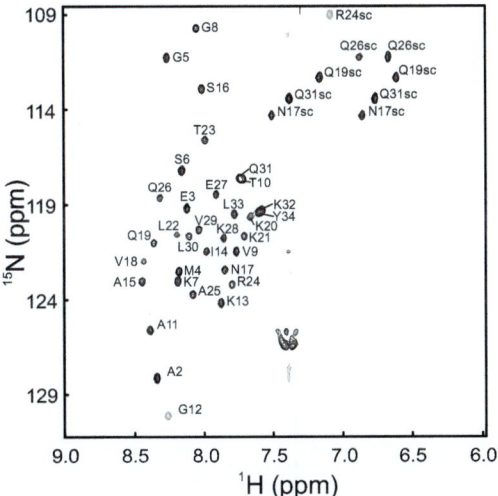

SI Figure 13. Relative cross peak intensity of N-BAR peptide backbone amide protons at 250 ms exchange time during the MEXICO experiment.

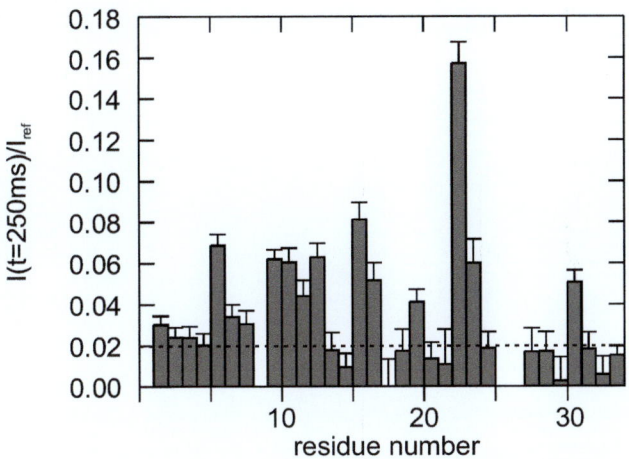

Danksagung

Zum Schluss möchte ich allen ganz herzlich danken, die an der Entstehung dieser Arbeit beteiligt waren:

Die vorliegende Arbeit wurde am Lehrstuhl für Biochemie an der Universität Bayreuth und am Lehrstuhl für Biophysik der Universität Halle-Wittenberg unter der Leitung von Prof. Dr. Jochen Balbach angefertigt. Ihm gilt mein besonderer Dank für die erstklassige wissenschaftliche Betreuung, die gewährte Freiheit und das Vertrauen bei der Entwicklung neuer Projekte. Er hat mir dadurch eine spannende, abwechslungs- und lehrreiche Doktorarbeitszeit ermöglicht, wofür ich sehr dankbar bin.

Ich möchte mich auch bei Prof. Dr. Franz X. Schmid für die stete Hilfsbereitschaft, Disskussionsfreude und zahlreichen guten Ratschläge bedanken.

Vielen Dank an meine Laborkolleginnen und –kollegen Barbara Eckert, Insa Kather, Christine Magg, Michael Wunderlich, Stefan Lorenz, Roman Jakob, Claudia Staab, Markus Zeeb, Ulrich Weininger, Michael Kovermann, Caroline Haupt, Rolf Sachs, Mohanraj Gopalswamy, Rica Patzschke, Stefan Gröger, Katrin Waldheim und Detlef Reichert für die gute Zusammenarbeit und hervorragende Laboratmosphäre.

Ein besonderer Dank gilt Ulrich Weininger. Uli, zusammen sind wir nach Halle gekommen und mussten uns dort erstmal zu Recht finden. Danke für die ausgezeichnete Zusammenarbeit und Schreibtischnachbarschaft! Nächtelanges Messen, Spektren analysieren, Projekte diskutieren, grillen in der Peißnitz, das machen wir mal wieder. Viel Glück für die Promotionsarbeit!

Bei Markus Zeeb bedanke ich mich für die Einführung in die Methoden der Proteinfaltung.

Vielen Dank auch an Piotr Neumann. Piotr, danke für deinen unermüdlichen Optimismus, Überzeugung, Einsatz und Enthusiasmus für die gemeinsamen Kristallisationsprojekte.

Nicht zu vergessen die ausschweifenden Diskussionen mit Hauke Lilie, Ralph Golbik und Hagen Hofmann über Sinn und Unsinn der Proteinfaltung. Herzlichen Dank dafür.

Ein Dankeschön auch an Hwankyu Lee, Prof. Dr. Richard W. Pastor, Ines Neundorf, Prof. Dr. Annette G. Beck-Sickinger, Kristian Schweimer, Nadine Homeyer, Prof. Dr. Heinrich Sticht, Mirjam Klepsch, Piotr Neumann, Beatrice Epler, Prof. Dr. Milton T. Stubbs, Wei Zhang, Prof. Dr. Ernest D. Laue, Henning Tidow, Cindy Schulenburg und Christian Scholz für die tollen Kooperationen im Laufe dieser Projekte.

Dem Graduiertenkolleg 1026 „Conformational transitions" unter der Leitung von Prof. Dr. Milton T. Stubbs und Mechthild Wahle (Koordinatorin des Graduiertenkollegs) danke ich für die tolle Zeit und freundliche Unterstützung während meiner Doktorandenzeit.

Ich möchte mich auch bei der Boehringer-Ingelheim- und GlaxoSmithKline-Stiftung, sowie dem Graduiertenkolleg 1026 für die finanzielle Unterstützung bedanken.

Und nicht zuletzt möchte ich meinen lieben Eltern danken! Danke, dass ihr an mich glaubt und mich immer untersützt. Die bayrischen Wurzeln nehme ich überallhin mit. Meinen Geschwistern Manfred und Kathrin danke ich, dass Sie mir –gerade während der letzten Jahre - den Rücken frei gehalten haben. Vielen Dank auch an Mirjam und meine Freunde, denn es gibt auch ein Leben neben der Wissenschaft.

Die VDM Verlagsservicegesellschaft sucht für wissenschaftliche Verlage abgeschlossene und herausragende

Dissertationen, Habilitationen, Diplomarbeiten, Master Theses, Magisterarbeiten usw.

für die kostenlose Publikation als Fachbuch.

Sie verfügen über eine Arbeit, die hohen inhaltlichen und formalen Ansprüchen genügt, und haben Interesse an einer honorarvergüteten Publikation?

Dann senden Sie bitte erste Informationen über sich und Ihre Arbeit per Email an *info@vdm-vsg.de*.

Sie erhalten kurzfristig unser Feedback!

VDM Verlagsservicegesellschaft mbH
Dudweiler Landstr. 99   Telefon +49 681 3720 174
D - 66123 Saarbrücken   Fax   +49 681 3720 1749
www.vdm-vsg.de

Die VDM Verlagsservicegesellschaft mbH vertritt

Printed by Books on Demand GmbH, Norderstedt / Germany